Defect Correction Methods for Fluid Flows at High Reynolds Numbers

Defect Correction Methods for Fluid Flows at High Reynolds Numbers presents the mathematical development of defect correction methods (DCM) in application to fluid flow problems in various settings. We will show several approaches to applying the DCM ideas in computational fluid dynamics (CFD) – from a basic idea of controlling the flow by the means of increased diffusion, to the state-of-the-art family of novel, DCM-based turbulence models. The main idea of the methods presented in this book, is to use defect correction in turbulence modelling; additionally, several methods will also be presented, that aim at reducing the time discretization error.

Features

- Provides a road map, starting from the ideas of minimally invasive controlling of turbulent flows, to the ways of improving the existing regularization techniques with DCM, to the ideas of 'full defect correction' in both space and time and, finally, to the more complex embedding of the DCM into turbulence modelling by the 'correction' of the whole turbulence model
- Can be used for teaching a topics course on a Masters or Ph.D. level. It is even more suitable as a reference for CFD theorists and practitioners, with most of the methods being minimally invasive and, therefore, easy to implement in the existing/legacy codes
- Discusses the current challenges in turbulence modelling with defect correction, showing several possible directions for future developments. Two source codes are provided – one for a regularization technique and another for a novel turbulence model – in order to give an interested researcher a quick start to the topic of DCM in CFD.

Alexander Labovsky is an Associate Professor of Mathematical Sciences at Michigan Technological University. His research interests include high accuracy methods for PDEs, multiscale modelling and numerical analysis of turbulence models for Navier-Stokes and coupled Navier-Stokes systems (fluid-fluid interaction, magnetohydrodynamics). He also has ongoing projects in Uncertainty Quantification and modelling of compressible flows.

Numerical Analysis and Scientific Computing Series

This series, comprising of a diverse collection of textbooks, references, and handbooks, brings together a wide range of topics across numerical analysis and scientific computing. The books contained in this series will appeal to an academic audience, both in mathematics and computer science, and naturally find applications in engineering and the physical sciences.

Modelling with Ordinary Differential Equations
A Comprehensive Approach
Alfio Borzì

Numerical Methods for Unsteady Compressible Flow Problems
Philipp Birken

A Gentle Introduction to Scientific Computing
Dan Stanescu, Long Lee

Introduction to Computational Engineering with MATLAB
Timothy Bower

An Introduction to Numerical Methods
A MATLAB® Approach, Fifth Edition
Abdelwahab Kharab, Ronald Guenther

The Sequential Quadratic Hamiltonian Method
Solving Optimal Control Problems
Alfio Borzì

Advances in Theoretical and Computational Fluid Mechanics
Existence, Blow-up, and Discrete Exterior Calculus Algorithms
Terry Moschandreou, Keith Afas, Khoa Nguyen

Stochastic Methods in Scientific Computing
From Foundations to Advanced Techniques
Massimo D'Elia, Kurt Langfeld, Biagio Lucini

Defect Correction Methods for Fluid Flows at High Reynolds Numbers
Alexander Labovsky

For more information about this series please visit: https://www.crcpress.com/Chapman--HallCRC-Numerical-Analysis-and-Scientific-Computing-Series/book-series/CHNUANSCCOM

Defect Correction Methods for Fluid Flows at High Reynolds Numbers

Alexander E. Labovsky

CRC Press
Taylor & Francis Group
Boca Raton London New York

CRC Press is an imprint of the
Taylor & Francis Group, an **informa** business
A CHAPMAN & HALL BOOK

Front cover image: Shutterstock

First edition published 2026
by CRC Press
2385 NW Executive Center Drive, Suite 320, Boca Raton FL 33431

and by CRC Press
4 Park Square, Milton Park, Abingdon, Oxon, OX14 4RN

CRC Press is an imprint of Taylor & Francis Group, LLC

ISBN: 978-1-032-41043-2 (hbk)
ISBN: 978-1-032-41041-8 (pbk)
ISBN: 978-1-003-35597-7 (ebk)

DOI: 10.1201/ 9781003355977

Typeset in CMR10 font
by KnowledgeWorks Global Ltd.

Publisher's note: This book has been prepared from camera-ready copy provided by the authors.

In memory of my father

Contents

Preface and Acknowledgments

When starting their research, most graduate students (at least in mathematics) tend to follow in the footsteps of their advisor. It helps a young mathematician tremendously if the advisor's area of interest is vast. I was lucky - my PhD advisor William Layton, just like his advisor Max Gunzburger, wears many hats. These include, among others, the computational fluid dynamics (CFD) and the predictor–corrector methods. The first project that Bill had suggested to me, was the regularization of a fluid flow via defect correction. In the following twenty years, I worked, among other projects, on various types of predictor–corrector ideas for different fluid flow systems. At some point, I got convinced that the bad reputation of the defect correction methods in the CFD community is a misunderstanding. As is often the case in mathematics, the CFD researchers started to slowly introduce the predictor–corrector methods into fluid flow modeling: initially, these were used to remedy the "insufficient" viscosity in steady flows, by adding hyperviscosity in the predictor step and removing it during the correction stage. It worked fairly well as a stabilization technique, but the expectations were higher. Having used the predictor–corrector methods in a similar manner (limiting them to changing the size of a specific term in the equation) for the incompressible and compressible fluid flows at increasing Reynolds numbers, the CFD researchers have largely dismissed defect correction as too crude of an idea for a topic as fragile as fluid flow modeling.

I will try to convince the readers that there is a flaw in this argument, and that the defect correction methods should be applied to the system as a whole, rather than affect one term in the equation. If implemented correctly, this allows the methods to be used in the very challenging topic of turbulence modeling.

Turbulent flows exhibit their most interesting behavior near a physical obstacle; coincidentally, in turbulence modeling almost everything feels like an obstacle. Every "common" step in fluid flow modeling - space discretization, time discretization, choosing an iterative solver and a preconditioner for the large matrix of the system - could fail in its own unique way. Even if the chosen turbulence model, the discretization, the iterative solver and the problem data seem to be working well for the chosen turbulent benchmark problem, the ice is always very thin: change the problem data (flow speed, viscosity, forcing, initial and boundary conditions) or move to a different benchmark problem - and the turbulence model is no longer as efficient as it seemed to be. This leads to an abundance of the existing turbulence models, and the skepticism of CFD researchers when they hear of yet another promising turbulence model appearing on the horizon.

I would like to convince the readers, who want to contribute to the field of turbulence modeling, to take a different route: instead of creating new models, let us improve the existing ones. The first step to take is to clarify, how we should quantify the improvement. The algorithm that I propose in the second half of this book (Chapters 7–11) aims to run the chosen turbulence model twice, in parallel, in a predictor–corrector fashion, in order to reduce parts of the overall error. This should not be viewed as a magic "fix" to all the shortcomings of the given model; rather, compared to an existing turbulence model, this build-up allows to reduce the required CPU time or to obtain a higher quality solution in a comparable amount of time. I call this the higher accuracy per computational cost.

Flow of the book

The book could be viewed as comprised of two parts. The first part, Chapters 2–6, examines various approaches to fluid flow modeling with defect correction. These include a simple regularization idea for the Navier–Stokes equations (NSE) in Chapter 2; a way of increasing the accuracy of a decoupled system of equations – magnetohydrodynamics (MHD) in Chapter 3 and a fluid–fluid interaction problem (FFI) in Chapter 6; and a way of improving the temporal accuracy of an existing turbulence model in Chapter 4. In Chapter 5, an attempt is made to replace a turbulence model with defect correction. The flow of these topics starts with the description of the method, followed by some theoretical results and, inevitably, numerical validation.

The second part of the book, Chapters 7–11, proposes and investigates the algorithm for reducing the overall error of existing turbulence models via defect correction. The idea, generally applicable in most of the existing approaches to turbulence modeling, is applied in this book only to the family of Large Eddy Simulation (LES) turbulence models. The emerging models, called LES-C (Large Eddy Simulation with Correction), are investigated in different settings, including the NSE, MHD and FFI. The results show the favorable comparison of the LES-C models to their natural competitors, the LES models. I hope that this new idea and the abundance of areas of application will appeal to graduate students and established researchers alike. Despite the high value of the Reynolds number in many of the test problems, the "classical" turbulent benchmark problems are not included in this book, making this an obvious idea for a challenging new project. Two source codes, one for a simple regularization of the NSE and one for the novel turbulence model, are included in this book to make the start of such projects as smooth as possible.

Acknowledgments

I would like to thank Callum Fraser, my editor at Taylor & Francis, for his guidance and especially for his patience. Much of this book is based on the papers I had written with my friends and collaborators, and I am thankfully listing them here: Mustafa Aggul, Yasasya Batugedara, Jeffrey Connors, Dilek Erkmen, Max Gunzburger, Eda Onal, Kyle Schwiebert, Catalin Trenchea, and Nick Wilson. I would especially like to thank my former PhD students Mustafa, Dilek, Yasasya, and Kyle for their belief in the LES-C idea in the early stages of the project and for the majority of the plots in the second part of this book.

This book is dedicated to my parents, Eugene and Larisa, for their unconditional support, for their love, and for the pride that they took in my relative success. The book is also dedicated to my wife, Anastasia, and my sons, Matthew and Maxwell, for the joy they bring into my life every day.

CHAPTER 1

Introduction

This book is a marriage proposal. By the end of Chapter 11 the author hopes to convince the reader that turbulence modeling should go hand in hand with defect correction. To quote Richard Feynman, turbulence is "the most important unsolved problem of classical physics". There are many areas where incompressible (discussed herein) and compressible turbulent flows can be of interest, including engineering, astrophysics, climate modeling, weather prediction, and (somewhat surprisingly) blood flows. There exists a vast body of literature on the topic of fluid flow modeling and turbulence modeling. We refer the interested reader to the book by Stephen Pope, [167], and the great variety of references therein. In this book we will investigate ways of incorporating defect correction approaches into the modeling of incompressible turbulence. Efforts have been made in the past to implement defect correction (predictor–corrector) schemes in turbulence modeling, but the results were not appealing enough to the Computational Fluid Dynamics (CFD) community.

1.1 TURBULENCE

Fluid flows, governed by the Navier–Stokes equations, can be characterized by many parameters, one of which is the Reynolds number. It is proportional to the flow's velocity and the reference length of the domain, and inverse proportional to the viscosity coefficient: $Re = LU/\nu$. In cases when the Reynolds number is large, the flow often enters the so-called turbulence regime. Such flows typically have sharp changes in velocity and pressure, and they are very hard to model. The term "turbulence" could be found in Leonardo da Vinci's "Codex Atlanticus", "...doue la turbolenza dellacqua si genera...", but it has probably been of interest long before Leonardo's time. Many brilliant minds have worked to deepen our understanding of turbulence, including L. Richardson (1920s), J. Leray, G.I. Taylor and Th. von Karman (1930s), A. Kolmogorov (1940s), and many others. Lewis Fry Richardson is the author of the famous phrase "Big whirls have little whirls that feed on their velocity; little whirls have lesser whirls and so on to viscosity", [2]. Jean Leray shaped up the idea of

DOI: 10.1201/9781003355977-1

turbulence modeling and proposed the first turbulence model in [6]. Inspired by the works of Taylor [1] and von Karman [3], Andrei Kolmogorov wrote a series of papers in the early 1940s, [92], commonly referred to as the K-41 theory, where he described in detail what is known as the energy cascade: the energy enters the turbulent systems at large scales, then the nonlinear (convection) term of the Navier–Stokes momentum equation acts to transport the energy through the so-called "inertial range" of scales at the fixed rate of $E(k) \simeq k^{-5/3}$ and the flow diffusion acts to dissipate this energy at the flow's microscale $\eta \simeq Re^{-3/4}$. Here k is a wavenumber and $E(k)$ represents the energy in the flow scales of the size $O(1/k)$.

This continuum of scales present in a turbulent flow, from the domain-size scales to the microscale, poses the main difficulty in turbulence modeling. In an airflow around a moving car the Reynolds number could reach $Re = 10^8$, and in a flow around the commercial airplane one can get $Re = 10^{12}$. Thus, resolving all scales up to the flow's microscale would be prohibitively expensive now and in the foreseeable future, even with supercomputers. This explains the need for turbulence modeling.

Per Wilcox [5], there exist so many turbulence models that no one model could be viewed as *the* turbulence model. This statement was made in 1998, and the number of turbulence models has increased even further in the last quarter of a century. This could be read as "instead of creating new models, try to improve the existing ones". One of the main ideas of this book, presented and discussed in Chapters 7–11, is to take a closer look at the existing turbulence models and find a way to reduce the modeling error (and the temporal discretization error, why not?) via defect correction.

1.2 DEFECT CORRECTION METHODS

Defect correction methods have been applied to ordinary and partial differential equations for many years – see, e.g., the paper by L. Fox from 1947, [4]. Generally speaking, the idea of a defect correction procedure is to use a chosen attractive numerical method (e.g., one that is stable and computationally inexpensive, albeit low order accurate) and improve its results – e.g., increase the accuracy of the method – via a correction step (or a sequence of such steps). The corrections are based on the residual error, which could stand for the overall error of the chosen method – or any part of the overall error. A survey of defect correction methods should include the works of Zadunaisky [7, 8] and Frank [9], as well as Frank and Ueberhuber [10], Stetter [101], and Böhmer, Hemker, and Stetter, [12]. Many references to defect correction in CFD will be given throughout this book. In particular, a subfamily of defect correction methods was proposed by Dutt et al. in [36]: the authors used the term "deferred correction methods", and these aim to reduce the error due to temporal discretization.

The chapters of this book discuss various defect correction methods for fluid flow problems – including the subfamily of deferred correction methods.

The methods will vary in complexity and goals (some of these were built as "regularization techniques", while others present themselves as potent enough to be called turbulence models). The applications will also vary – from the Navier–Stokes equations (NSE) to coupled NS systems, including Magnetohydrodynamics and a fluid–fluid interaction problem.

The reader is encouraged to use the techniques discussed in this book, to create and implement new methods. Source code will be provided in Chapters 2 and 7 to help interested readers develop their own codes.

CHAPTER 2

Artificial Viscosity and Defect Correction

2.1 A SIMPLE IDEA FOR DEFECT CORRECTION

Incompressible fluid flows are governed by the following system of the time-dependent, incompressible Navier–Stokes equations (NSE):

$$\begin{aligned} \frac{\partial u}{\partial t} - Re^{-1}\Delta u + u \cdot \nabla u + \nabla p &= f, \text{ for } x \in \Omega, 0 < t \leq T, \qquad (2.1) \\ \nabla \cdot u &= 0, x \in \Omega, 0 < t \leq T, \\ u(x,0) &= u_0(x), x \in \Omega, \end{aligned}$$

where $\Omega \subset \mathbb{R}^d, d = 2, 3$. Boundary conditions should be imposed on $\partial\Omega$ for $0 < t \leq T$.

We begin our journey with one of the simpler regularization ideas for the Navier–Stokes equations. The first defect correction method of this book will be based on the concept of artificial viscosity: the model's kinematic viscosity is increased, adding numerical stability at the cost of the reduced accuracy of the solution. Defect correction is then employed to correct for the addition of the non-physical viscosity.

Let a kth order accurate discretization of the *equilibrium* NSE be written as

$$NSE^h(u^h) = f, \qquad (2.2)$$

The k-step DCM computes $u_1^h, ..., u_k^h$ as

$$\begin{aligned} -\alpha h \Delta^h u_1^h + NSE^h(u_1^h) &= f, \qquad (2.3) \\ -\alpha h \Delta^h u_l^h + NSE^h(u_l^h) &= f - \alpha h \Delta^h u_{l-1}^h, \text{ for } l = 2, ..., k, \end{aligned}$$

where the velocity approximations u_i^h are sought in the finite element space of piecewise polynomials of degree k.

DOI: 10.1201/9781003355977-2

For many years, it has been widely believed that (2.3) can be directly imported into implicit time discretizations of flow problems in the obvious way: discretize in time, given $u^h(t_{OLD})$, the quasistatic flow problem for $u^h(t_{NEW})$ is solved by applying (2.3) directly, resulting in

$$-\alpha h\Delta^h u_1^h(t_{NEW}) + B(u_1^h(t_{NEW}), u_k^h(t_{OLD})) + NSE^h(u_1^h(t_{NEW})) = f, \quad (2.4)$$
$$-\alpha h\Delta^h u_l^h(t_{NEW}) + B(u_l^h(t_{NEW}), u_k^h(t_{OLD})) + NSE^h(u_l^h(t_{NEW}))$$
$$= f - \alpha h\Delta^h u_{l-1}^h(t_{NEW}), \text{ for } l = 2, ..., k,$$

where B is a time-stepping operator (e.g., Backward Euler).

It seems reasonable to use the "best" approximation available for $u(t_{OLD})$ – that is, $u_k^h(t_{OLD})$. However, this natural idea doesn't seem to be even stable (see the discussion at the end of this chapter).

Denote the velocity and pressure spaces by X and Q, respectively. Let $X^h \subset X, Q^h \subset Q$ be finite dimensional, conforming finite element spaces.

The Backward Euler time discretization, combined with the two-step defect correction method in space, leads to the following system of equations for $(u_1^{h,n+1}, p_1^{h,n+1})$, $(u_2^{h,n+1}, p_2^{h,n+1}) \in (X^h, Q^h), \forall v^h \in X^h$ at $t = t_{n+1}$, $n \geq 0$, with $\Delta t = t_{i+1} - t_i$

$$\left(\frac{u_1^{h,n+1} - u_1^{h,n}}{\Delta t}, v^h\right) + (h + Re^{-1})(\nabla u_1^{h,n+1}, \nabla v^h) \quad (2.5)$$
$$+ b^*(u_1^{h,n+1}, u_1^{h,n+1}, v^h) - (p_1^{h,n+1}, \nabla \cdot v^h) = (f(t_{n+1}), v^h),$$
$$\left(\frac{u_2^{h,n+1} - u_2^{h,n}}{\Delta t}, v^h\right) + (h + Re^{-1})(\nabla u_2^{h,n+1}, \nabla v^h) + b^*(u_2^{h,n+1}, u_2^{h,n+1}, v^h)$$
$$- (p_2^{h,n+1}, \nabla \cdot v^h) = (f(t_{n+1}), v^h) + h(\nabla u_1^{h,n+1}, \nabla v^h),$$

where $b^*(\cdot,\cdot,\cdot)$ is the explicitly skew-symmetrized trilinear form,

$$b^*(u, v, w) := \frac{1}{2}(u \cdot \nabla v, w) - \frac{1}{2}(u \cdot \nabla w, v).$$

Notice the structure of the defect and correction step equations – first and second equations of (4.4). The left-hand sides of both equations are identical, and the extra terms added to the right-hand side of the correction equation depend solely upon the defect solution u_1^h. Thus, the method is naturally parallelizable, as the solutions $u_1^{h,n+1}$ and $u_2^{h,n}$ can be sought simultaneously, at two cores, and at the same amount of CPU time.

2.2 NUMERICAL ANALYSIS

The following stability results were proven in [11].

Lemma 1 (Stability of the AV approximation). *Let u_1^h satisfy the first equation of (4.4). Let $f \in L^2(0,T; H^{-1}(\Omega))$. Then for $n = 0, ..., N-1$*

$$\|u_1^{h,n+1}\|^2 + \Delta t\Sigma_{i=1}^{n+1}(h + Re^{-1})\|\nabla u_1^{h,i}\|^2 \leq \|u_0\|^2$$

$$+\frac{1}{h+Re^{-1}}\Delta t\Sigma_{i=1}^{n+1}\|f(t_i)\|_{-1}^2.$$

Also, if $f \in L^2(0,T;L^2(\Omega))$ and the time constraint T is finite, then there exists a constant $C = C(T)$ such that

$$\begin{aligned}\|u_1^{h,n+1}\|^2 + \Delta t\Sigma_{i=1}^{n+1}(h+Re^{-1})\|\nabla u_1^{h,i}\|^2 \\ \leq C(\|u_0\|^2 + \Delta t\Sigma_{i=1}^{n+1}\|f(t_i)\|^2).\end{aligned} \tag{2.6}$$

Theorem 1 (Stability). *Let u_1^h, u_2^h satisfy (4.4). Let $f \in L^2(0,T;H^{-1}(\Omega))$. Then for $n = 0,...,N-1$: $u_1^{h,n+1}, u_2^{h,n+1}$ are bounded and*

$$\begin{aligned}\|u_2^{h,n+1}\|^2 + \frac{2h^2}{(h+Re^{-1})^2}\|u_1^{h,n+1}\|^2 + \Delta t\Sigma_{i=1}^{n+1}(h+Re^{-1})\|\nabla u_2^{h,i}\|^2 \\ \leq \left(1+\frac{2h^2}{(h+Re^{-1})^2}\right)\|u_0\|^2 \\ +\left(1+\frac{h^2}{(h+Re^{-1})^2}\right)\frac{2}{h+Re^{-1}}\Delta t\Sigma_{i=1}^{n+1}\|f(t_i)\|_{-1}^2.\end{aligned} \tag{2.7}$$

Also, if $f \in L^2(0,T;L^2(\Omega))$ and the time constraint T is finite, then there exists a constant $C = C(T)$ such that

$$\begin{aligned}\|u_2^{h,n+1}\|^2 + \frac{2h^2}{(h+Re^{-1})^2}\|u_1^{h,n+1}\|^2 + \Delta t\Sigma_{i=1}^{n+1}(h+Re^{-1})\|\nabla u_2^{h,i}\|^2 \\ \leq C(\|u_0\|^2 + \Delta t\Sigma_{i=1}^{n+1}\|f(t_i)\|^2).\end{aligned} \tag{2.8}$$

It follows from (2.8) that both approximations u_1^h and u_2^h are bounded at any time level and for any viscosity, provided that the initial approximation and the forcing term are L^2-integrable.

2.2.1 Error Estimates

The defect correction method in (4.4) introduces a "defect" of the size $O(h)$, and corrects for it. Thus, the general theory of the defect correction methods suggests the following error convergence rates in approximating the NSE velocity u by the defect step approximation u_1 and the correction step approximation u_2:

$$\begin{aligned}\|u-u_1^h\|_{energy-norm} \leq C(h^m + h + \Delta t^s), \\ \|u-u_2^h\|_{energy-norm} \leq C(h^m + h^2 + \Delta t^s).\end{aligned} \tag{2.9}$$

Here the velocity approximations u_1^h and u_2^h are sought in the finite-element space of piecewise polynomials of degree m, and we assume that a time discretization method is chosen, with the global error of size $O(\Delta t^s)$. For the Backward Euler time discretization, used in (4.4), we have $s = 1$.

TABLE 2.1 Errors and Convergence Rates, Defect Step. $Re = 100,000$

h	$\|\|u - u_1^h\|\|_{L^2(0,T;L^2(\Omega))}$	rate	$\|\|u - u_1^h\|\|_{L^2(0,T;H^1(\Omega))}$	rate
1/4	0.000339015	-	0.00534596	-
1/8	7.39569E-05	2.2	0.00104601	2.35
1/16	3.19763E-05	1.21	0.00025783	2.02
1/32	1.62156E-05	0.98	9.19028E-05	1.49

Theorems 2–5 in [11] prove that the estimates (2.9) hold if the true solution is smooth enough. The proof of accuracy of the correction step solution requires the "intermediate" result: the accuracy of the time derivative $\frac{e_1^{n+1}-e_1^n}{\Delta t}$ must be proven, where $e_1^i = u(t_i) - u_1^{h,i}$.

2.3 COMPUTATIONAL TESTS

2.3.1 Convergence Rates

A simple two-dimensional test problem can illustrate the advantage of the correction step solution. Consider the Chorin's vortex decay problem in the unit square $\Omega = (0,1)^2$. Take the true solution

$$u = \begin{pmatrix} -\cos(\pi x)\sin(\pi y)exp(-2\pi^2 t/Re) \\ \sin(\pi x)\cos(\pi y)exp(-2\pi^2 t/Re) \end{pmatrix}, \tag{2.10}$$

$$p = -\frac{1}{4}(\cos(2\pi x) + \cos(2\pi y))exp(-4\pi^2 t/Re),$$

and compute the right-hand side f and initial condition u_0, so that (2.10) satisfies (2.1).

In order to ensure that the dominating part of the overall error is the modeling error, the time step is taken to be very small, $\Delta t = O(h^3)$. The following results were obtained for the case of high Reynolds number [11].

TABLE 2.2 Errors and Convergence Rates, Correction Step. $Re = 100,000$

h	$\|\|u - u_2^h\|\|_{L^2(0,T;L^2(\Omega))}$	rate	$\|\|u - u_2^h\|\|_{L^2(0,T;H^1(\Omega))}$	rate
1/4	0.000300427	-	0.00525358	-
1/8	0.000040032	2.91	0.000975526	2.43
1/16	5.94795E-06	2.75	0.000190267	2.36
1/32	1.37357E-06	2.11	4.26364E-05	2.16

There are two important takeaway messages from these tables. First, the claimed convergence rates are verified for this test problem, and the convergence of the correction step solution is indeed quadratic, whereas the defect step solution converges linearly. Secondly, comparing the errors in both solutions, we notice that a prescribed tolerance could be reached by the correction step solution u_2^h on a coarser mesh than needed for the defect step solution u_1^h.

2.3.2 Comparison of the Approaches – Should We use the Best Available Solution?

Now we turn to the important question: shouldn't the best available approximation be used in the chosen time marching scheme? In other words, shouldn't the realization of the DCM for the time dependent NSE result in (2.4)? Specifically, for the formulation (4.4), we would have $u_2^{h,n}$ replacing $u_1^{h,n}$ in the first term of the defect step equation.

In order to test both alternatives, we use the one-dimensional singularly perturbed equation

$$u_t - \epsilon u_{xx} + u_x = f$$

The DCM used in (4.4) leads to the coupled pair of equations

$$\frac{u_{1,i}^{n+1} - u_{1,i}^n}{\Delta t} - (\epsilon + h)\frac{u_{1,i-1}^{n+1} - 2u_{1,i}^{n+1} + u_{1,i+1}^{n+1}}{h^2} + \frac{u_{1,i+1}^{n+1} - u_{1,i-1}^{n+1}}{2h} = f_i^{n+1},$$

$$\frac{u_{2,i}^{n+1} - u_{2,i}^n}{\Delta t} - (\epsilon + h)\frac{u_{2,i-1}^{n+1} - 2u_{2,i}^{n+1} + u_{2,i+1}^{n+1}}{h^2} + \frac{u_{2,i+1}^{n+1} - u_{2,i-1}^{n+1}}{2h} = f_i^{n+1} - h\frac{u_{1,i-1}^{n+1} - 2u_{1,i}^{n+1} + u_{1,i+1}^{n+1}}{h^2},$$

whereas the alternative being tested is

$$\frac{u_{1,i}^{n+1} - u_{2,i}^n}{\Delta t} - (\epsilon + h)\frac{u_{1,i-1}^{n+1} - 2u_{1,i}^{n+1} + u_{1,i+1}^{n+1}}{h^2} + \frac{u_{1,i+1}^{n+1} - u_{1,i-1}^{n+1}}{2h} = f_i^{n+1},$$

$$\frac{u_{2,i}^{n+1} - u_{2,i}^n}{\Delta t} - (\epsilon + h)\frac{u_{2,i-1}^{n+1} - 2u_{2,i}^{n+1} + u_{2,i+1}^{n+1}}{h^2} + \frac{u_{2,i+1}^{n+1} - u_{2,i-1}^{n+1}}{2h} = f_i^{n+1} - h\frac{u_{1,i-1}^{n+1} - 2u_{1,i}^{n+1} + u_{1,i+1}^{n+1}}{h^2}.$$

Consider $0 \le x \le 1$, $0 \le t \le 1$, $u(0,t) = 0$, $u(1,t) = 25$, $u(x,0) = 0$, $f(x,t) = tx^2 + 20x$, $\epsilon = 0.000001$.

Figures 2.1–2.4 show the solution, obtained by the DCM (denoted by the solid line) and the solution, obtained by the alternative approach (dashed line on the graphs). The spacial mesh is fixed at $h = 0.01$ and the time step Δt decreases to zero (see the captions).

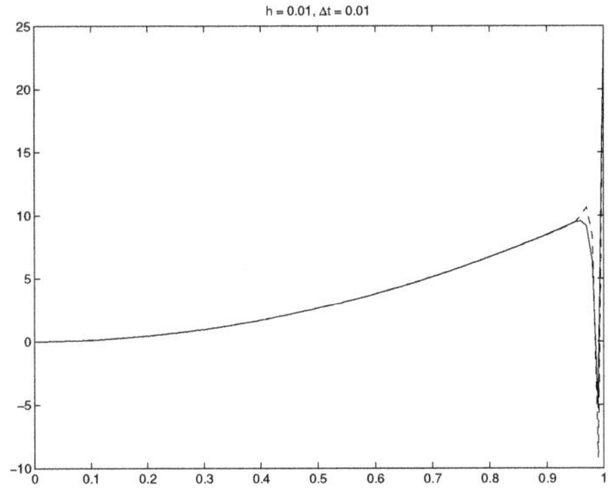

Figure 2.1 $\Delta t = 0.01.$

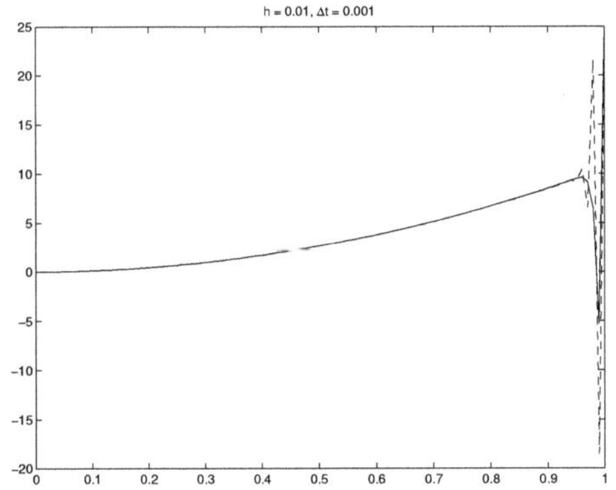

Figure 2.2 $\Delta t = 0.001.$

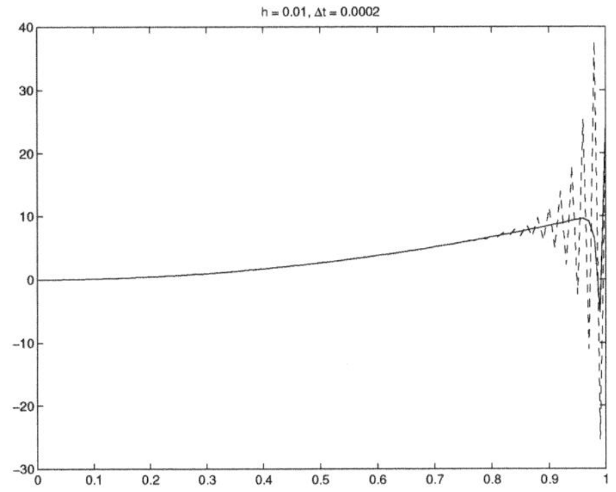

Figure 2.3 $\Delta t = 0.0002.$

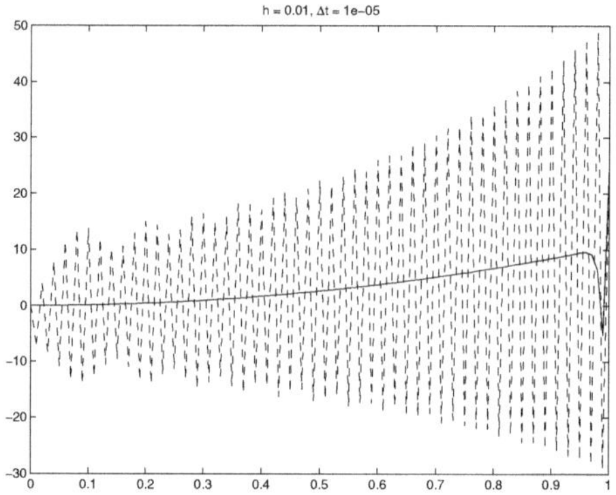

Figure 2.4 $\Delta t = 0.00001$.

As we see, although this alternative approach uses the Correction Step approximation of the true solution on each time level (instead of the AV approximation), the results are worse even for a simple one-dimensional problem with the bounded domain and bounded right-hand side.

We conclude the comparison of the methods by applying them to the Navier–Stokes equations in $\mathbb{R}^2$. Consider the Chorin's vortex decay problem in the square $\Omega = (-\frac{1}{2}, \frac{1}{2})^2$ with

$$f(x,y,t) = \begin{pmatrix} \frac{1}{2}\pi\sin(2\pi x)exp(-4\pi^2 t/Re) \\ \frac{1}{2}\pi\sin(2\pi y)exp(-4\pi^2 t/Re) \end{pmatrix} \tag{2.11}$$

and $Re = 10^5$. The final time is taken to be $T = 1/10$ and the mesh diameter is fixed at $h = 1/4$. For such a coarse mesh, having a very small time step means that the overall error is not affected by the time discretization error, and the error estimates obtained by the DCM (4.4) do not change, as the time step Δt is decreased, Table 2.3.

Table 2.4 shows the results of the alternative approach.

In the alternative approach the error *increases* as Δt tends to zero. The two examples, given above, were convincing enough for the author to never use the alternative approach in defect correction. In all the methods described

TABLE 2.3 DCM

Δt	$\|\| u - u_2^h \|\|_{L^2(0,T;L^2(\Omega))}$	$\|\| u - u_2^h \|\|_{L^2(0,T;H^1(\Omega))}$
10^{-3}	0.00682	0.0585
10^{-4}	0.00682	0.0585
10^{-5}	0.00682	0.0585

$Re = 100,000$, $T = 1/10$, $h = 1/4$

TABLE 2.4 Alternative Approach

Δt	$\|\|u-u_2^h\|\|_{L^2(0,T;L^2(\Omega))}$	$\|\|u-u_2^h\|\|_{L^2(0,T;H^1(\Omega))}$
10^{-3}	0.01019	0.1104
10^{-4}	0.01449	0.1759
10^{-5}	0.01582	0.2076

$Re = 100,000$, $T = 1/10$, $h = 1/4$

in this book, the left-hand sides in the i-th equation of a k-step DCM ($i < k$), solving for u_i^{n+1} at the time level $(n+1)$, will contain u_i^n and not u_k^n.

2.4 SOURCE CODE

Presented below is the FreeFEM++ [124] code that was used in the convergence rates test in this chapter.

```
//   Backward Euler for the time-dependent NSE. Defect correction
     method\index{defect correction} (artificial viscosity
    approximation and the correction for it).

real s0=clock();
border A(t=0,1){x=t; y=0; label=1;};
border B(t=0,1){x=1; y=t; label=2;};
border C(t=0,1){x=1-t; y=1; label=3;};
border D(t=0,1){x=0; y=1-t; label=4;};

int N=32; // number of mesh points on one axis - change this to
    refine the mesh
real h=1.0/N; // meshsize

mesh Th=square(N, N);

fespace Vh2(Th,P2);
fespace Vh(Th,P1);

Vh2 u2,v2,u2old,um2,e2, bd1, bd2;
Vh2 u1,v1,u1old,um1,e1;
Vh2 dcvel1, dcvel2, dcvel1old, dcvel2old, dcvelm1, dcvelm2, dce1,
     dce2;
Vh  p, q, e3, pold, dcp, dcpold, dce3;


real reynolds=100000.;
real nu = 1.0/reynolds;
real err1, err2, dcerr1;
real errL2 = 0.;
real errL2L2 = 0.;
real currenterrL2 = 0.;
real dcerrL2 = 0.;
```

```
real dcerrL2L2 = 0.;
real dccurrenterrL2 = 0.;

real errH1 = 0.;
real errL2H1 = 0.;
real currenterrH1 = 0.;
real dcerrH1 = 0.;
real dcerrL2H1 = 0.;
real dccurrenterrH1 = 0.;


real dt=0.1*h*h*h;
int nbiter = N*N*N/32.0;
int itnl=0,iter=0, MaxNlIt=30;
real t=0;

func real truevel1 (real t) {
return -cos(pi*x)*sin(pi*y)*exp(-2*pi^2*t*nu);
}

func real truevel2 (real t) {
return sin(pi*x)*cos(pi*y)*exp(-2*pi^2*t*nu);
}

func real truepres (real t) {
return (-1.0/4)*(cos(2*pi*x) + cos(2*pi*y))*exp(-4*pi^2*t*nu);
}

func real dxtruevel1 (real t) {
return pi*sin(pi*x)*sin(pi*y)*exp(-2*pi^2*t*nu);
}

func real dxtruevel2 (real t) {
return pi*cos(pi*x)*cos(pi*y)*exp(-2*pi^2*t*nu);
}

func real dytruevel1 (real t) {
return -pi*cos(pi*x)*cos(pi*y)*exp(-2*pi^2*t*nu);
}

func real dytruevel2 (real t) {
return -pi*sin(pi*x)*sin(pi*y)*exp(-2*pi^2*t*nu);
}


u1 = truevel1(0);
u2 = truevel2(0);
p = truepres(0);

dcvel1 = truevel1(0);
dcvel2 = truevel2(0);
dcp = truepres(0);

u1old=0;
u2old=0;
um1=0;
um2=0;
```

```
pold=0;

dcvel1old=0;
dcvel2old=0;
dcvelm1=0;
dcvelm2=0;
dcpold=0;

problem NSDCM1 ([u1,u2,p],[v1,v2,q],solver=UMFPACK) =
    int2d(Th)(
              (1.0/dt)*( u1*v1 + u2*v2)

            + (u1old*dx(u1) + u2old*dy(u1))*v1
            + (u1old*dx(u2) + u2old*dy(u2))*v2

            + (u1*dx(u1old) + u2*dy(u1old))*v1
            + (u1*dx(u2old) + u2*dy(u2old))*v2

            + (h + nu) * ( dx(u1)*dx(v1) + dy(u1)*dy(v1)
            +         dx(u2)*dx(v2) + dy(u2)*dy(v2) )

            + p*q*(0.000000001)

            - p*(dx(v1) + dy(v2))

            - q*(dx(u1) + dy(u2))
           )
  + int2d(Th) (
            - (1.0/dt)*( um1*v1 + um2*v2)

            - (u1old*dx(u1old) + u2old*dy(u1old))*v1
            - (u1old*dx(u2old) + u2old*dy(u2old))*v2
             )
  + on(1,2,3,4, u1=bd1,u2=bd2);

problem NSDCM2 ([dcvel1,dcvel2,dcp],[v1,v2,q],solver=UMFPACK) =
    int2d(Th)(
              (1.0/dt)*( dcvel1*v1 + dcvel2*v2)

            + (dcvel1old*dx(dcvel1) + dcvel2old*dy(dcvel1))*v1
            + (dcvel1old*dx(dcvel2) + dcvel2old*dy(dcvel2))*v2

            + (dcvel1*dx(dcvel1old) + dcvel2*dy(dcvel1old))*v1
            + (dcvel1*dx(dcvel2old) + dcvel2*dy(dcvel2old))*v2

            + (h + nu) * ( dx(dcvel1)*dx(v1) + dy(dcvel1)*dy(v1)
            +         dx(dcvel2)*dx(v2) + dy(dcvel2)*dy(v2) )

            + dcp*q*(0.000000001)

            - dcp*(dx(v1) + dy(v2))

            - q*(dx(dcvel1) + dy(dcvel2))
           )
  + int2d(Th) (
            - (1.0/dt)*( dcvelm1*v1 + dcvelm2*v2)
```

```
              - (dcvel1old*dx(dcvel1old) + dcvel2old*dy(dcvel1old))
                  *v1
              - (dcvel1old*dx(dcvel2old) + dcvel2old*dy(dcvel2old))
                  *v2

              - h * ( dx(u1)*dx(v1) + dy(u1)*dy(v1)
              +         dx(u2)*dx(v2) + dy(u2)*dy(v2) )
                )
  + on(1,2,3,4, dcvel1=bd1, dcvel2=bd2);


for (iter=1;iter<=nbiter;iter++)
{
   um1 = u1;
   um2 = u2;
   u1old = u1;
   u2old = u2;
   pold = p;

   dcvelm1 = dcvel1;
   dcvelm2 = dcvel2;
   dcvel1old = dcvel1;
   dcvel2old = dcvel2;
   dcpold = dcp;

   t = t + dt;

   bd1 = truevel1(t);
   bd2 = truevel2(t);

   err1 = 1;
   dcerr1 = 1;
   cout << " iter = " << iter << " ------------------------ " <<
       endl;
//
   for (itnl=1; itnl<=MaxNlIt; itnl++)
   {
       cout<< "itnl = " << itnl << endl ;
       NSDCM1;
       e1 = abs(u1 - u1old);
       e2 = abs(u2 - u2old);

       cout<< "e1[].max " << e1[].max << "      e2[].max " << e2
           [].max<< "e3[].max " << e3[].max << endl ;
       cout<< " " << endl ;
       if ( (e1[].max < 0.0001) && (e2[].max < 0.0001))
        {
           break ;
        }
        u1old = u1 ;
        u2old = u2 ;
        err1 = err1 + 1 ;
   }

   if ( err1 >= MaxNlIt )
   {
      cout << "approx failed for iter = " << iter << endl ;
```

```
      break;
  }

  //————————————————————————————

  currenterrL2 = int2d(Th)((truevel1(t) - u1)^2 + (truevel2(t) -
        u2)^2);
  errL2 = errL2 + currenterrL2;

  currenterrH1 = int2d(Th)((dxtruevel1(t) - dx(u1))^2 + (
      dxtruevel2(t) - dx(u2))^2 + (dytruevel1(t) - dy(u1))^2 + (
      dytruevel2(t) - dy(u2))^2);
  errH1 = errH1 + currenterrH1;

//  ----------------Now the Correction
    step-----------------------------------------------

  for (itnl=1; itnl<=MaxNlIt; itnl++)
  {
      cout<< "itnl = " << itnl << endl ;
      NSDCM2;
      dce1 = abs(dcvel1 - dcvel1old);
      dce2 = abs(dcvel2 - dcvel2old);

      cout<< "dce1[].max " << dce1[].max << "       dce2[].max "
          << dce2[].max<< "          dce3[].max " << dce3[].max <<
          endl ;
      cout<< " " << endl ;
      if ( (dce1[].max < 0.0001) && (dce2[].max < 0.0001))
       {
          break ;
       }
       dcvel1old = dcvel1 ;
       dcvel2old = dcvel2 ;
       dcerr1 = dcerr1 + 1 ;
  }

  if ( err1 >= MaxNlIt )
  {
     cout << "approx failed for iter = " << iter << endl ;
     break;
  }

  //————————————————————————————

  dccurrenterrL2 = int2d(Th)((truevel1(t) - dcvel1)^2 + (
      truevel2(t) - dcvel2)^2);
  dcerrL2 = dcerrL2 + dccurrenterrL2;

  dccurrenterrH1 = int2d(Th)((dxtruevel1(t) - dx(dcvel1))^2 + (
      dxtruevel2(t) - dx(dcvel2))^2 + (dytruevel1(t) - dy(dcvel1)
      )^2 + (dytruevel2(t) - dy(dcvel2))^2);
  dcerrH1 = dcerrH1 + dccurrenterrH1;

}
  errL2L2 = sqrt(dt*errL2);
  dcerrL2L2 = sqrt(dt*dcerrL2);
```

```
    errL2H1 = sqrt(dt*(h + nu)*errH1);
    dcerrL2H1 = sqrt(dt*(h + nu)*dcerrH1);


cout << "||u1 - true||_L2-L2 = " << errL2L2 << endl;
cout << "||u2 - true||_L2-L2 = " << dcerrL2L2 << endl;
cout << " " << endl;
cout << "||u1 - true||_L2-H1 = " << errL2H1 << endl;
cout << "||u2 - true||_L2-H1 = " << dcerrL2H1 << endl;

cout << "CPU " << clock()-s0 << "s " << endl;
```

CHAPTER 3

Deferred Correction in MagnetoHydroDynamics

3.1 DECOUPLING THE MHD SYSTEM

Let us turn our attention to coupled Navier–Stokes systems. In many such problems, including fluid–fluid interactions, fluid-structure ineractions, climate modeling and Magnetohydrodynamics (MHD), one aims to decouple the problem. This allows for more efficient modeling (solving two systems with N degrees of freedom is more preferable than solving one $2N$-sized system, especially if it could be done in parallel), and it could sometimes also allow for the usage of the pre-existing legacy codes, written for the parts of the current problem. Such a decoupling is highly nontrivial, as the methods should be stable and efficient enough. We will discuss the challenges in decoupling a fluid–fluid interaction problem in the later chapters; now let us turn our attention to the modeling of MHD flows.

When an electrically conducting fluid moves in a magnetic field, the mathematical equations that describe its motion are known as the MHD system. If the fluid flow is presumed incompressible, the MHD equations (see [44, 56]) are

$$\frac{\partial \boldsymbol{u}}{\partial t} + (\boldsymbol{u} \cdot \nabla)\boldsymbol{u} - (\boldsymbol{B} \cdot \nabla)\boldsymbol{B} - \nu \Delta \boldsymbol{u} + \nabla p = \boldsymbol{f}, \tag{3.1}$$

$$\frac{\partial \boldsymbol{B}}{\partial t} + (\boldsymbol{u} \cdot \nabla)\boldsymbol{B} - (\boldsymbol{B} \cdot \nabla)\boldsymbol{u} - \nu_m \Delta \boldsymbol{B} = \nabla \times \boldsymbol{g}, \tag{3.2}$$

$$\nabla \cdot \boldsymbol{u} = 0, \tag{3.3}$$

$$\nabla \cdot \boldsymbol{B} = 0. \tag{3.4}$$

An excellent detailed derivation of (3.1)–(3.4) from the Navier–Stokes and Maxwell's equations is given in [44], including the discussion on the coupling of the velocity field $\boldsymbol{u}$ and magnetic field $\boldsymbol{B}$ via the Lorenz force and Ohm's law.

DOI: 10.1201/9781003355977-3

In Chapter 10, we will revisit the MHD system (3.1)–(3.4) when discussing the MHD turbulence modeling; there we will assume that the kinematic viscosity ν and magnetic diffusivity ν_m are very small, in which case the flow may enter the turbulent regime.

Remark 1. *The system* (3.1)−(3.4) *has more equations than it has unknowns. The divergence free form of the right hand side in* (3.2) *guarantees that the magnetic field* $\boldsymbol{B}$ *is divergence free, provided that the initial condition satisfies this constraint,* $\nabla \cdot \boldsymbol{B}(0, \mathbf{x}) = 0$*. Thus, equation* (3.4) *can be omitted. An equivalent formulation of the MHD system can be written, with a general (not necessarily divergence free) forcing g in the right-hand side in* (3.2)*; a Lagrange multiplier* $\nabla\lambda$ *would then need to be added to the left-hand side of* (3.2).

One of the main difficulties in solving the system (3.1)–(3.4) is its size: six dimensions must be resolved to find the 3-D velocity and magnetic fields. With the inherent difficulties of the Navier–Stokes systems magnified in the MHD case, this puts additional strain on the computational resources. So, as an alternative to the monolithic (implicit, fully coupled) methods, that are robust but very computationally expensive, researchers sought partitioned approaches, where the MHD equations are decoupled. Not only would these methods reduce the computational cost substantially (solving two N-by-N systems, instead of a $2N$-by-$2N$ system), but they could potentially allow for the usage of legacy codes.

3.1.1 Elsässer Variables and the IMEX Method

One approach to decoupling the MHD system in a stable way was presented in [73]. It starts with the splitting of the magnetic field in two parts $\boldsymbol{B} = \boldsymbol{B}_\circ + \boldsymbol{b}$ (mean and fluctuations).

Then MHD system is re-written in the Elsässer variables [49]

$$\boldsymbol{z}^+ = \boldsymbol{u} + \boldsymbol{b}, \quad \boldsymbol{z}^- = \boldsymbol{u} - \boldsymbol{b}, \tag{3.5}$$

allowing for a stable time-splitting scheme. A Lagrange multiplier is added to the left-hand side of (3.2), and the requirement that the forcing in (3.2) must be solenoidal, is waived. The momentum equations, in the Elsässer variables, are

$$\frac{\partial \boldsymbol{z}^\pm}{\partial t} \mp (\boldsymbol{B}_\circ \cdot \nabla)\boldsymbol{z}^\pm + (\boldsymbol{z}^\mp \cdot \nabla)\boldsymbol{z}^\pm - \frac{\nu + \nu_m}{2}\Delta \boldsymbol{z}^\pm - \frac{\nu - \nu_m}{2}\Delta \boldsymbol{z}^\mp + \nabla p^\pm = f^\pm, \tag{3.6}$$

while the continuity equations are $\nabla \cdot \boldsymbol{z}^\pm = 0$. We note that the nonlinear interactions occur between the Alfvènic fluctuations $\boldsymbol{z}^\pm$. The authors show in [73] that the following IMEX (implicit–explicit) decoupling method is unconditionally stable. The solution pair $(\boldsymbol{z}, p)$ of (3.6) is approximated by the pair $(\tilde{\boldsymbol{z}}, \tilde{p})$ satisfying

$$
\begin{aligned}
&\frac{\tilde{\boldsymbol{z}}^{\pm}_{n+1} - \tilde{\boldsymbol{z}}^{\pm}_{n}}{\Delta t} \mp (\boldsymbol{B}_{\circ} \cdot \nabla)\tilde{\boldsymbol{z}}^{\pm}_{n+1} + (\tilde{\boldsymbol{z}}^{\mp}_{n} \cdot \nabla)\tilde{\boldsymbol{z}}^{\pm}_{n+1} && (3.7)\\
&\qquad - \frac{\nu + \nu_m}{2}\Delta\tilde{\boldsymbol{z}}^{\pm}_{n+1} - \frac{\nu - \nu_m}{2}\Delta\tilde{\boldsymbol{z}}^{\mp}_{n} + \nabla\tilde{p}^{\pm}_{n+1} = \boldsymbol{f}^{\pm}(t_{n+1}),\\
&\nabla \cdot \tilde{\boldsymbol{z}}^{\pm}_{n+1} = 0. && (3.8)
\end{aligned}
$$

The structure of the MHD-IMEX equations (3.7)–(3.8) resembles that of the Navier–Stokes momentum equations. This allows for the existing (legacy) codes to be used, which is especially valuable when modeling MHD flows in complex geometries. This makes the method even more appealing, leaving only one serious flaw: the low order of temporal accuracy.

3.2 REDUCING THE TIME DISCRETIZATION ERROR VIA DEFERRED CORRECTION

The decoupling mechanism at the heart of the MHD-IMEX system is not readily extendable to the "usual" higher order time marching methods, like the multistep or Runge–Kutta methods. However, the deferred correction procedure seems to be quite suitable to unblock this bottleneck. Deferred correction methods for ODEs, introduced by Dutt et al. in the seminal paper [36] and further developed by Minion et al. (see, e.g., [91]), have been briefly discussed in Chapter 1.

The authors in [76] used the MHD-IMEX as the defect step and built a (deferred) correction step to obtain the following system for the variables $(\tilde{\tilde{\boldsymbol{z}}}^{\pm}_{n+1}, \tilde{\tilde{p}}^{\pm})$:

$$
\begin{aligned}
&\frac{\tilde{\tilde{\boldsymbol{z}}}^{\pm}_{n+1} - \tilde{\tilde{\boldsymbol{z}}}^{\pm}_{n}}{\Delta t} - \frac{\nu + \nu_m}{2}\Delta\tilde{\tilde{\boldsymbol{z}}}^{\pm}_{n+1} - \frac{\nu - \nu_m}{2}\Delta\tilde{\tilde{\boldsymbol{z}}}^{\pm}_{n} \mp (\boldsymbol{B}_{\circ} \cdot \nabla)\tilde{\tilde{\boldsymbol{z}}}^{\pm}_{n+1} && (3.9)\\
&+ (\tilde{\tilde{\boldsymbol{z}}}^{\mp}_{n} \cdot \nabla)\tilde{\tilde{\boldsymbol{z}}}^{\pm}_{n+1} + \nabla\tilde{\tilde{p}}^{\pm}_{n+1} = -\frac{\nu + \nu_m}{2}\Delta\tilde{\boldsymbol{z}}^{\pm}_{n+1} + \frac{\nu + \nu_m}{4}\Delta\tilde{\boldsymbol{z}}^{\pm}_{n+1} + \frac{\nu + \nu_m}{4}\Delta\tilde{\boldsymbol{z}}^{\pm}_{n}\\
&- \frac{\nu - \nu_m}{2}\Delta\tilde{\boldsymbol{z}}^{\mp}_{n} + \frac{\nu - \nu_m}{4}\Delta\tilde{\boldsymbol{z}}^{\pm}_{n+1} + \frac{\nu - \nu_m}{4}\Delta\tilde{\boldsymbol{z}}^{\pm}_{n}\\
&\mp (\boldsymbol{B}_{\circ} \cdot \nabla)\tilde{\boldsymbol{z}}^{\pm}_{n+1} \pm \frac{1}{2}(\boldsymbol{B}_{\circ} \cdot \nabla)\tilde{\boldsymbol{z}}^{\pm}_{n+1} \pm \frac{1}{2}(\boldsymbol{B}_{\circ} \cdot \nabla)\tilde{\boldsymbol{z}}^{\pm}_{n}\\
&- \frac{1}{2}(\tilde{\boldsymbol{z}}^{\mp}_{n+1} \cdot \nabla)\tilde{\boldsymbol{z}}^{\pm}_{n+1} - \frac{1}{2}(\tilde{\boldsymbol{z}}^{\mp}_{n+1} \cdot \nabla)\tilde{\boldsymbol{z}}^{\pm}_{n}\\
&- \nabla\tilde{p}^{\pm}_{n+1} + \frac{1}{2}\nabla(\tilde{p}^{\pm}_{n+1} + \tilde{p}^{\pm}_{n})\\
&+ \frac{1}{2}\boldsymbol{f}(t_{n+1})^{\pm} + \frac{1}{2}\boldsymbol{f}(t_{n})^{\pm},\\
&\nabla \cdot \tilde{\tilde{\boldsymbol{z}}}^{\pm}_{n+1} = 0. && (3.10)
\end{aligned}
$$

We intentionally leave the terms in the right-hand side of (3.9) non simplified, to illustrate the idea of spectral deferred correction [36, 91]: all but the time derivative terms in the left-hand side of the correction step are added also to the right-hand side, but the quantity of interest is taken at the previous iteration of the deferred correction procedure (in our case, it means that the defect solutions $\mathbf{z}, p$ are used instead of the correction step solutions $\tilde{\mathbf{z}}, \tilde{p}$). Then a time average of these terms, over the current time step, is subtracted from the right hand side, with the trapezoid rule used for the time integral quadrature.

Both pairs $(\tilde{z}, \tilde{p})$ and $(\tilde{\tilde{z}}, \tilde{\tilde{p}})$ are approximations to the solution (z, p) of (3.6). Interestingly, if one replaces both pairs $(\tilde{z}, \tilde{p})$ and $(\tilde{\tilde{z}}, \tilde{\tilde{p}})$ in (3.9) with the pair (z, p), one obtains the second order accurate Crank–Nicolson method for the coupled (monolithic) system (3.6). This gives one way of intuitively understanding the increased accuracy of the correction step (3.9). Note that the defect and correction steps can – and should – be run in parallel on two processor cores in a trapezoidal fashion, with the pairs $(\tilde{z}^{\pm}_{n+1}, \tilde{p}^{\pm}_{n+1})$ and $(\tilde{\tilde{z}}^{\pm}_{n}, \tilde{\tilde{p}}^{\pm}_{n})$ computed simultaneously.

3.3 THEORETICAL RESULTS

Both the defect step and the correction step solutions were shown in [76] to be unconditionally stable; in general, the stability properties of the correction step solutions in a deferred correction procedure are similar to the stability properties of the defect step solutions. The interesting question is the claimed improvement in the accuracy. Does the correction step provide the second order accurate solution, despite having the same IMEX structure in the left-hand side of the system? The answer is given in the two theorems from [76], that prove the accuracy of the two approximate solutions. The IMEX method, used in the defect step, is shown to be first order accurate in time.

Theorem 2. *Given a final time $T > 0$ and timestep $\Delta t > 0$ let $t_n = n \times \Delta t$ for $n = 0, 1, ..., N$. Let $\tilde{z}^{\pm}_{i}$ denote the MHD-IMEX approximation of $z^{\pm}(t_i)$, $i = 0, 1, ..., N$, and let*
$\tilde{e}^{\pm}_{i} := z^{\pm}(t_i) - \tilde{z}^{\pm}_{i}$. Then provided the true solution of (3.6) satisfies the regularity assumptions

$$\begin{aligned}
\nabla z^{\pm} &\in L^2((0,T); L^{\infty}(\Omega)), \\
\partial_t z^{\pm} &\in L^2((0,T); L^2(\Omega)), \\
\partial_{tt} z^{\pm} &\in L^2((0,T); L^2(\Omega)), \\
\nabla \partial_t z^{\pm} &\in L^2((0,T); L^2(\Omega)),
\end{aligned}$$

and for a set of constants $\eta_i \in (t_i, t_{i+1})$, $i = 0, 1, ..., N$, *the following error estimate holds*

$$\begin{aligned}
&\|\tilde{e}_N^+\|^2 + \|\tilde{e}_N^-\|^2 + \Delta t \frac{\nu\nu_m}{2(\nu+\nu_m)} \sum_{n=0}^{N} (\|\nabla \tilde{e}_n^+\|^2 + \|\nabla \tilde{e}_n^-\|^2) \\
&\quad \leq \Delta t^2 C \frac{\nu+\nu_m}{\nu\nu_m} \exp\left(\Delta t \sum_{n=0}^{N-1} (\|\nabla \boldsymbol{z}^-(t_n)\|^2 + \|\nabla \boldsymbol{z}^+(t_n)\|^2\right) \times \\
&\quad \Delta t \sum_{n=0}^{N} \Big(\|\partial_{tt} \boldsymbol{z}^+(\eta_n)\|^2 + \|\partial_t \boldsymbol{z}^-(\eta_n)\|^2 \|\nabla \boldsymbol{z}^+(t_n)\|_\infty^2 \\
&\qquad + (\nu_m - \nu)^2 (\|\nabla \partial_t \boldsymbol{z}^-(\eta_n)\|^2 + \|\nabla \partial_t \boldsymbol{z}^+(\eta_n)\|^2) \\
&\qquad + \|\partial_{tt} \boldsymbol{z}^-(\eta_n)\|^2 + \|\partial_t \boldsymbol{z}^+(\eta_n)\|^2 \|\nabla \boldsymbol{z}^-(t_n)\|_\infty^2 \Big).
\end{aligned} \tag{3.11}$$

As is often the case with defect correction methods in CFD, the accuracy of the correction step requires an intermediate result, where the accuracy of (higher) derivatives must be established. For the correction step (3.9) we need the accuracy of the time difference of the error in the defect step approximation. This can be seen if the like terms are combined in the right-hand side of (3.9). For instance, the first three terms simplify to

$$\begin{aligned}
&-\frac{\nu+\nu_m}{2} \Delta \tilde{\boldsymbol{z}}_{n+1}^\pm + \frac{\nu+\nu_m}{4} \Delta \tilde{\boldsymbol{z}}_{n+1}^\pm + \frac{\nu+\nu_m}{4} \Delta \tilde{\boldsymbol{z}}_n^\pm \\
&\qquad = -\frac{\nu+\nu_m}{4} \Delta t \left(\frac{\Delta \tilde{\boldsymbol{z}}_{n+1}^\pm - \Delta \tilde{\boldsymbol{z}}_n^\pm}{\Delta t} \right).
\end{aligned}$$

In the error analysis these terms are subtracted from the corresponding terms containing the true solution. Thus, in order for such terms to produce higher temporal accuracy, an accuracy result must be obtained for $\frac{\tilde{\boldsymbol{e}}_{i+1}^\pm - \tilde{\boldsymbol{e}}_i^\pm}{\Delta t}$, where $\tilde{\boldsymbol{e}}_i^\pm := \boldsymbol{z}^\pm(t_i) - \tilde{\boldsymbol{z}}_i^\pm$. The result of Theorem 2, along with some extra requirements on the regularity of the true solution, suffices to show the first order accuracy of $\frac{\tilde{\boldsymbol{e}}_{i+1}^\pm - \tilde{\boldsymbol{e}}_i^\pm}{\Delta t}$, see the theorem below.

Theorem 3. *Given a final time* $T > 0$ *and timestep* $\Delta t > 0$ *let* $\tilde{\boldsymbol{z}}_n^\pm$ *be the IMEX approximation to* $\boldsymbol{z}(t_n)^\pm$ *for* $n = 1, 2, ..., N$. *Let* $\tilde{\boldsymbol{e}}_n^\pm = \boldsymbol{z}(t_n)^\pm - \tilde{\boldsymbol{z}}_n^\pm$. *Provided the true solution satisfies the additional regularity assumptions*

$$\begin{aligned}
\partial_{ttt} \boldsymbol{z}^\pm &\in L^2((0,T); L^2(\Omega)), \\
\partial_t \boldsymbol{z}^\pm &\in L^2((0,T); L^\infty(\Omega)), \\
\nabla \partial_t \boldsymbol{z}^\pm &\in L^2((0,T); L^\infty(\Omega))
\end{aligned}$$

the discrete time derivative of the IMEX is first order accurate in time and satisfies

$$\begin{aligned}
&\|\frac{\tilde{e}_N^+ - \tilde{e}_{N-1}^+}{\Delta t}\|^2 + \|\frac{\tilde{e}_N^- - \tilde{e}_{N-1}^-}{\Delta t}\|^2 \\
&+ \Delta t \frac{\nu\nu_m}{\nu+\nu_m} \sum_{n=1}^{N} \left(\left\| \frac{\nabla(\tilde{e}_n^+ - \tilde{e}_{n-1}^+)}{\Delta t} \right\|^2 + \left\| \frac{\nabla(\tilde{e}_n^- - \tilde{e}_{n-1}^-)}{\Delta t} \right\|^2 \right) \\
&\le C \frac{\nu+\nu_m}{\nu\nu_m} \exp\left(\Delta t \sum_{n=0}^{N-1} (\|\nabla z^+(t_n)\|_\infty^2 + \|\nabla z^-(t_{n+1})\|_\infty^2) \right) \times \\
&\left(\Delta t \sum_{n=0}^{N-1} \left(\Delta t^2 \|\partial_{ttt} z^+(\eta_n)\|^2 + \Delta t^2 \|\partial_{ttt} z^-(\eta_n)\|^2 + \Delta t^2 \|\nabla \partial_t z^+(\eta_n)\|_\infty^2 \right.\right. \\
&\quad + \Delta t^2 \|\nabla \partial_t z^-(\eta_n)\|^2 \|\nabla \partial_t z^+(\eta_n)\|^2 + \Delta t^2 \|\nabla \partial_{tt} z^-(\eta_n)\|^2 \|\nabla z^+(t_n)\|^2 \\
&\quad + \Delta t^2 \|\nabla \partial_{tt} z^+(\eta_n)\|^2 \|\nabla \partial_t z^-(t_n)\|^2 + \|\partial_t z^-(\eta_n)\|_\infty^2 \|\nabla \tilde{e}_{n+1}^+\|^2 \\
&\quad + \|\nabla \partial_t z^+(\eta_n)\|^2 \|\nabla \tilde{e}_n^-\|^2 + \|\partial_t z^+(\eta_n)\|_\infty^2 \|\nabla \tilde{e}_{n+1}^-\|^2 \\
&\quad \left.\left. + \|\nabla \partial_t z^-(\eta_n)\|^2 \|\nabla \tilde{e}_n^+\|^2 + \|\nabla \tilde{e}_n^+\|^2 + \|\nabla \tilde{e}_n^-\|^2 \right) \right)
\end{aligned} \tag{3.12}$$

Now the stage is set for the proof of higher accuracy of the correction step approximation. Under the assumptions of Theorems 2–3 (plus some more regularity assumptions) the following theorem states that the correction step solution is second order accurate in time.

Theorem 4. *Given final time $T > 0$ and timestep $\Delta t > 0$, let $\tilde{z}_n^\pm$ and $\tilde{\tilde{z}}_n^\pm$ respectively denote the defect step and correction step approximations to $z^\pm$ at time $t_n = n\Delta t$ for $n = 1, 2, ..., N$, and let $\tilde{\tilde{e}}_n^\pm := z^\pm(t_n) - \tilde{\tilde{z}}_n^\pm$. Additionally, we let $s_{n+1}^\pm = \frac{1}{\Delta t}(\tilde{e}_{n+1}^\pm - \tilde{e}_n^\pm)$ for $n = 1, 2, ..., N-1$. Provided the true solution satisfies the additional regularity*

$$\begin{aligned}
&\partial_{tt}(z^\mp \cdot \nabla) z^\pm \in L^2((0,T); L^2(\Omega)), \\
&\partial_{tt} \nabla z^\pm \in L^2((0,T); L^2(\Omega)), \ \textit{and} \\
&\partial_{tt} f^+, \partial_{tt} f^- \in L^2((0,T); L^2(\Omega)),
\end{aligned}$$

then the correction step approximation is second order accurate in time and satisfies

$$
\begin{aligned}
&\|\tilde{\tilde{e}}_N^+\|^2 + \|\tilde{\tilde{e}}_N^-\|^2 + \Delta t \frac{\nu\nu_m}{\nu+\nu_m} \sum_{n=0}^{N} (\|\nabla \tilde{\tilde{e}}_n^+\|^2 + \|\nabla \tilde{\tilde{e}}_n^-\|^2) \\
&\quad \le C\frac{\nu+\nu_m}{\nu\nu_m} \exp\left(\Delta t \sum_{n=0}^{N-1} \|\nabla \boldsymbol{z}^+(t_n)\|_\infty^2 + \|\nabla \boldsymbol{z}^-(t_n)\|_\infty^2\right) \times \\
&\quad \left(\Delta t \sum_{n=0}^{N-1} C\frac{\nu+\nu_m}{\nu\nu_m} \{\Delta t^2 \|\nabla \boldsymbol{B}_\circ\|^2 \|\nabla \boldsymbol{s}_{n+1}^+\|^2 + \Delta t^2 \|\nabla \boldsymbol{B}_\circ\|^2 \|\nabla \boldsymbol{s}_{n+1}^-\|^2\right. \\
&\qquad + \Delta t^2 \|\nabla \boldsymbol{z}^-(t_n)\|^2 \|\nabla \boldsymbol{s}_{n+1}^+\|^2 + \Delta t^2 \|\nabla \tilde{e}_n^-\|^2 \|\nabla \boldsymbol{s}_{n+1}^+\|^2 \\
&\qquad + \Delta t^2 \|\nabla \tilde{\boldsymbol{z}}_n^-\|^2 \|\nabla \boldsymbol{s}_{n+1}^+\|^2 + \Delta t^2 \|\nabla \boldsymbol{s}_{n+1}^-\|^2 \|\nabla \tilde{e}_{n+1}^+\|^2 \\
&\qquad + \Delta t^2 \|\nabla \partial_t \boldsymbol{z}^-(\eta_n)\|^2 \|\nabla \tilde{e}_{n+1}^+\|^2 + \Delta t^2 \|\nabla \boldsymbol{s}_{n+1}^-\|^2 \|\nabla \boldsymbol{z}^+(t_{n+1})\|^2 \\
&\qquad + \Delta t^2 \|\nabla \tilde{e}_{n+1}^-\|^2 \|\nabla \partial_t \boldsymbol{z}^+(\eta_n)\|^2 + \Delta t^2 \|\nabla \boldsymbol{s}_{n+1}^-\|^2 \|\nabla \boldsymbol{z}^+(t_n)\|^2 \\
&\qquad + \Delta t^2 \|\nabla \boldsymbol{z}^+(t_n)\|^2 \|\nabla \boldsymbol{s}_{n+1}^-\|^2 + \Delta t^2 \|\nabla \tilde{e}_n^+\|^2 \|\nabla \boldsymbol{s}_{n+1}^-\|^2 \\
&\qquad + \Delta t^2 \|\nabla \tilde{\boldsymbol{z}}_n^+\|^2 \|\nabla \boldsymbol{s}_{n+1}^-\|^2 + \Delta t^2 \|\nabla \boldsymbol{s}_{n+1}^+\|^2 \|\nabla \tilde{e}_{n+1}^-\|^2 \\
&\qquad + \Delta t^2 \|\nabla \partial_t \boldsymbol{z}^+(\eta_n)\|^2 \|\nabla \tilde{e}_{n+1}^-\|^2 + \Delta t^2 \|\nabla \boldsymbol{s}_{n+1}^+\|^2 \|\nabla \boldsymbol{z}^-(t_{n+1})\|^2 \\
&\qquad + \Delta t^2 \|\nabla \tilde{e}_{n+1}^+\|^2 \|\nabla \partial_t \boldsymbol{z}^-(\eta_n)\|^2 + \Delta t^2 \|\nabla \boldsymbol{s}_{n+1}^+\|^2 \|\nabla \boldsymbol{z}^-(t_n)\|^2 \\
&\qquad + \Delta t^2 (\nu+\nu_m)^2 \|\nabla \boldsymbol{s}_{n+1}^+\|^2 + \Delta t^2 (\nu-\nu_m)^2 \|\nabla \boldsymbol{s}_{n+1}^-\|^2 \\
&\qquad + \Delta t^2 (\nu+\nu_m)^2 \|\nabla \boldsymbol{s}_{n+1}^-\|^2 + \Delta t^2 (\nu-\nu_m)^2 \|\nabla \boldsymbol{s}_{n+1}^+\|^2 \\
&\qquad \left. + \Delta t^4 \|\partial_{tt} \boldsymbol{F}(\eta_n)\|^2 + \Delta t^4 \|\partial_{tt} \boldsymbol{F}(\eta_n)\|^2\} \right).
\end{aligned}
\tag{3.13}
$$

3.4 COMPUTATIONAL RESULTS – CONVERGENCE RATES

All the theoretical results in this chapter were given in the semi-continuous case: the solutions were discrete in time, but continuous in space. For the numerical validation of the theory, we turn to the fully discrete solution, with a finite element method used for spatial discretization.

Since we are mainly concerned with the temporal component of the overall error, we choose the two-dimensional wave propagation test problem: the solutions are varying smoothly in space so that it is easier to track the error due to the temporal discretization. In [76] an electrically conducting flow is considered, and the true solution (in Elsässer variables $\boldsymbol{z}^+, \boldsymbol{z}^-$) is chosen as follows.

TABLE 3.1 Accuracy of the Defect Step Solution

h	$\|z^+(T) - \tilde{z}_N^+\|_{L^2(\Omega)}$	*rate*	$\|z^-(T) - \tilde{z}_N^-\|_{L^2(\Omega)}$	*rate*
1/4	0.0260319		0.0240726	
1/16	0.0026907	1.64	0.00276714	1.56
1/64	0.000416796	1.35	0.000423207	1.35

$Re = 1$, $Re_m = 10$, $T = 1$.

$$\begin{cases} z^+ = \begin{pmatrix} 0.75 + 0.25\cos(2\pi(x-t))\sin(2\pi(y-t))e^{-8\pi^2 tRe^{-1}} + 0.1(y+1)^3 e^{tRe_m^{-1}} \\ -0.25\sin(2\pi(x-t))\cos(2\pi(y-t))e^{-8\pi^2 tRe^{-1}} + 0.1(x+1)^3 e^{tRe_m^{-1}} \end{pmatrix}, \\ z^- = \begin{pmatrix} 0.75 + 0.25\cos(2\pi(x-t))\sin(2\pi(y-t))e^{-8\pi^2 tRe^{-1}} - 0.1(y+1)^3 e^{tRe_m^{-1}} \\ -0.25\sin(2\pi(x-t))\cos(2\pi(y-t))e^{-8\pi^2 tRe^{-1}} - 0.1(x+1)^3 e^{tRe_m^{-1}} \end{pmatrix}, \\ p = -\frac{1}{64}(\cos(4\pi(x-t)) + \cos(4\pi(y-t)))e^{-16\pi^2 tRe^{-1}}, \end{cases}$$

with $Re = 1, Re_m = 10$.

The Taylor-Hood finite element pair (piecewise quadratic polynomials for z^+ and z^-, and piecewise linear polynomials for p) is chosen, so the theory predicts that the overall error is $O(h^2 + \Delta t)$ for the defect step and $O(h^2 + \Delta t^2)$ for the correction step. This should be true for both the $L^2((0,T); L^2(\Omega))$ and $L^2((0,T); H^1(\Omega))$ norms, but we only present the results for the $L^2(\Omega)$ norm at the final time $T = 1$, computed in [76].

Set the time step equal to the mesh size, $\Delta t = h$, to verify the claimed second-order accuracy of the method. The tables below demonstrate the first-order accuracy of the IMEX approximation $\tilde{z}$ and the second-order accuracy of the correction step approximation $\tilde{\tilde{z}}$.

The method (3.7) – (3.9) needs to be investigated further - at least computationally. It could be of interest to view separately the errors in the magnetic field and the velocity field approximation. Tables 3.1–3.2 show that the errors in the correction step are approximately 3.5 times smaller than those in the defect step; it would also be interesting to see the qualitative effects of the

TABLE 3.2 Accuracy of the Correction Step Solution

h	$\|z^+(T) - \tilde{\tilde{z}}_N^+\|_{L^2(\Omega)}$	*rate*	$\|z^-(T) - \tilde{\tilde{z}}_N^-\|_{L^2(\Omega)}$	*rate*
1/4	0.0444342		0.0466917	
1/16	0.00234416	2.12	0.00221543	2.2
1/64	0.000123658	2.12	0.00012288	2.09

$Re = 1$, $Re_m = 10$, $T = 1$.

correction step. To that end, one would choose a MHD benchmark problem with known – and subtle – qualitative features of the true solution, that would be hard to capture with the numerically efficient but not very accurate IMEX method (3.7); then apply the correction step (3.9) to see whether the quality of the solution is improved.

CHAPTER 4

Deferred Correction with a Turbulence Model

Starting with this chapter, a large portion of this book will be devoted to applying defect and deferred correction ideas to flows at high Reynolds numbers. Such flows often enter the so-called turbulent regime, characterized by sporadic changes in fluid velocity and pressure. Turbulent fluid flows are ubiquitous in nature, including atmospheric and ocean currents, flows through turbines and pipes, air flows past a moving car or plane, blood flow in arteries, etc. Accurate prediction of turbulent flows is important and sometimes crucial for human life; yet, as follows from the Kolmogorov K41 theory [92], the complexity of these flows makes the direct numerical simulation prohibitively expensive (now and in the foreseeable future). Thus, the only viable approach to studying turbulence flows is through modeling.

A large body of literature is available on the concept of turbulence and on turbulence modeling, from the seminal works of Richardson, Leray, Kolmogorov, to the dozens of new papers on turbulence appearing every month. In fact, there exist so many turbulence models that, clearly, no one model/approach can be viewed as the ultimate success.

In the chapters to follow, we will develop and investigate approaches to use defect correction on top of, instead of, and finally along with the existing turbulence models. We start with a simple idea, that deferred correction could be used to improve the accuracy of a time marching method, that is combined with a turbulence model for flows at high Reynolds numbers. Resolving turbulent flows, even with a turbulence model in place, is always computationally challenging, and any effort to reduce the computational time and/or the required computational resources, is of high demand. A parallelized deferred correction method could be advantageous, as it would allow for a low order, stable and computationally attractive time discretization method to be used, while still yielding a higher accuracy solution - all at a reasonable price of having to use more than one processor (core). We emphasize that, out of the

DOI: 10.1201/9781003355977-4

three major components of the overall error in the (fully discretized) case of turbulence modeling – the modeling error, the spatial discretization error, and the temporal discretization error – the method of this chapter aims to reduce only one component.

4.1 TURBULENCE MODELING

Flows at high Reynolds numbers cannot be resolved in the manner of direct numerical simulation. The need for modeling is justified by the works of Kolmogorov, who proved, [92], that there exists a continuum of scales in turbulent fluid flows, with the smallest scales (in the case of a $3-D$ flow) being of the order $O(Re^{-3/4})$, where the Reynolds number Re is inverse proportional to the viscosity coefficient ν. Thus, capturing all the small structures in a turbulent flow requires the number of mesh points in space for each time step to be $O(Re^{9/4})$ for three-dimensional problems. It is not uncommon to have $Re \sim O(10^{12})$ in real-life applications.

One of the popular approaches to turbulence modeling is known as Large Eddy Simulation. It is based on the idea that the flow can be represented by a collection of scales with different sizes, and the largest structures in the flow (containing most of the energy) are responsible for much of the mixing and most of the momentum transport. So, instead of trying to approximate all of them down to the smallest one, one defines a filter width $\delta > 0$ and computes only the scales of size larger than δ (large scales), whereas the effect of the small scales on the large scales is modeled. There exists a large body of literature on the topic of LES turbulence modeling; we refer the interested reader to [93, 94, 95] and the references therein.

If $\overline{(\cdot)}^{\delta}$ denotes a local, spatial averaging operator that commutes with differentiation, then averaging the NSE gives the following non-closed equations for the filtered quantities $\overline{u}^{\delta}$ and $\overline{p}^{\delta}$ in $(0,T)\times\Omega$:

$$\overline{u}^{\delta}_t + \nabla\cdot(\overline{u}^{\delta}\overline{u^T}^{\delta}) - \nu\Delta\overline{u}^{\delta} + \nabla\overline{p}^{\delta} + \nabla\cdot(\overline{uu^T}^{\delta} - \overline{u}^{\delta}\overline{u^T}^{\delta}) = \overline{f}^{\delta}, \tag{4.1a}$$

$$\nabla\cdot\overline{u}^{\delta} = 0. \tag{4.1b}$$

Throughout this book, we will consider the Germano differential filter, [79]. Given $\phi(x)$ in Ω, the filter A_{δ}^{-1} produces the unique solution $A_{\delta}^{-1}\phi := \overline{\phi}$ of $\overline{\phi} - \delta^2\Delta\overline{\phi} = \phi$ with zero or periodic boundary conditions. Note that under zero or periodic boundary conditions, this averaging operator commutes with differentiation.

4.1.1 Approximate Deconvolution Models

An LES model arises when one tries to close the system (4.1) by choosing an approximation to the last term on the left-hand side of (4.1a). Many different LES regularizations have been proposed and studied; we consider the family of Approximate Deconvolution Models (ADMs) that theoretically allow for arbitrarily high accuracy w.r.t. (with respect to) the filtering width. These models were introduced by Stolz and Adams in [80] and extensively studied; see, e.g., [81, 82, 83, 84, 85, 87]. Full numerical analysis is available for the ADM, along with the similarity theory [77]; we also refer an interested reader to the book [78].

The truncation of scales in the ADM is done with the aid of an approximate deconvolution operator - an approximate inverse to the chosen filter A_δ^{-1}.

Definition 1 (Approximate Deconvolution Operator). *For a fixed finite N, define the Nth approximate deconvolution operator G_N by*

$$G_N\phi = \sum_{n=0}^{N}(I - A_\delta^{-1})^n\phi,$$

The increased accuracy of the ADMs with respect to the filtering width stems from the following lemma, proven in [87].

Lemma 2. *The operator G_N^i is compact, positive, and is an asymptotic inverse to the filter A_δ^{-1}, i.e., for very smooth ϕ and as $\delta \to 0$, it satisfies*

$$\phi = G_N\overline{\phi}^\delta + (-1)^{N+1}\delta^{2N+2}\Delta^{N+1}A_\delta^{-(N+1)}\phi. \tag{4.2}$$

The velocity–pressure spaces considered in this chapter are

$$\begin{aligned} X &= H^1_{per}(\Omega)^d \\ &= \{v \in L^2(\Omega)^d : \nabla v \in L^2(\Omega)^{d\times d} \text{ and } v \text{ is periodic with period } L\}, \\ Q &= L^2_{per}(\Omega) = \{q : q \in L^2(\Omega), \int_\Omega q(x)dx = 0, q \text{ periodic with period } L\}. \end{aligned}$$

The finite dimensional, conforming finite element spaces, used for spatial discretization, are denoted by X^h, Q^h.

We will consider the simplest ADM ($N = 0$) which is second order accurate w.r.t. the filtering width δ, per Lemma 2. In the semi-continuous (continuous in time, discretized in space) setting the model seeks $(w^h, \zeta^h, q^h) \in (X^h, X^h, Q^h)$ such that for any $(v^h, \xi^h, \chi^h) \in (X^h, X^h, Q^h)$

$$\begin{aligned} (w_t^h, v^h) + \delta^2(\nabla w_t^h, \nabla v^h) + \nu(\nabla w^h, \nabla v^h) + \nu\delta^2(\nabla\zeta^h, \nabla v^h) \\ +b^*(w^h, w^h; v^h) - (q^h, \nabla\cdot v^h) = (f, v^h), \\ (\nabla w^h, \nabla\xi^h) = (\zeta^h, \xi^h), \\ (\nabla\cdot w^h, \chi^h) = 0. \end{aligned} \tag{4.3}$$

There is an extra variable ζ^h in this so-called mixed variational formulation (4.3). It is introduced so that the fourth order term $\nu\delta^2(\Delta w, \Delta v)$ does not prohibit the usage of C^0 finite elements; more details on this can be found in [57]. Notice that the ADM seeks approximation to the pair $(\bar{u}, p)$ of filtered velocity and non-filtered pressure.

4.2 ADM WITH DEFERRED CORRECTION

A deferred correction idea can be used here to reduce the second component of the overall error – the error due to temporal discretization. Note the second time dependent term in the left-hand side of (4.3). It poses extra difficulty in applying the deferred correction technique, which had been developed for ODEs of the form $\phi_t = F(t, \phi(t))$. The authors of [57] used a "straightforward" approach, applying the deferred correction to $A\phi_t = F(t, \phi(t))$, where A is the exact deconvolution, $A\bar{\phi} = \phi$. Starting with a low order, Backward Euler time discretization in the defect step, the correction step seeks a solution that is second order accurate in time.

The two-step deferred correction method computes (w_1^h, q_1^h) and (w_2^h, q_2^h), two consecutive approximations for the averaged velocity and pressure $(\bar{u}, p)$. These approximations satisfy the following equations for $(w_1^{h,n+1}, \zeta_1^{h,n+1}, q_1^{h,n+1})$, $(w_2^{h,n+1}, \zeta_2^{h,n+1}, q_2^{h,n+1}) \in (X^h, X^h, Q^h), \forall (v^h, \xi^h, \chi^h) \in (X^h, X^h, Q^h)$ at $t = t_{n+1}$, $n \geq 0$, with $k := \Delta t = t_{i+1} - t_i$:

$$
\begin{aligned}
&\left(\frac{w_1^{h,n+1} - w_1^{h,n}}{k}, v^h\right) + \delta^2\left(\nabla\left(\frac{w_1^{h,n+1} - w_1^{h,n}}{k}\right), \nabla v^h\right) + \nu(\nabla w_1^{h,n+1}, \nabla v^h) \\
&+ \nu\delta^2(\nabla \zeta_1^{h,n+1}, \nabla v^h) + b^*(w_1^{h,n+1}, w_1^{h,n+1}, v^h) - (q_1^{h,n+1}, \nabla \cdot v^h) \\
&\quad = (f(t_{n+1}), v^h),
\end{aligned} \tag{4.4a}
$$

$$
\begin{aligned}
&\left(\frac{w_2^{h,n+1} - w_2^{h,n}}{k}, v^h\right) + \delta^2\left(\nabla\left(\frac{w_2^{h,n+1} - w_2^{h,n}}{k}\right), \nabla v^h\right) + \nu(\nabla w_2^{h,n+1}, \nabla v^h) \\
&+ \nu\delta^2(\nabla \zeta_2^{h,n+1}, \nabla v^h) + b^*(w_2^{h,n+1}, w_2^{h,n+1}, v^h) - (q_2^{h,n+1}, \nabla \cdot v^h) \\
&\quad = \left(\frac{f(t_{n+1}) + f(t_n)}{2}, v^h\right) + \frac{\nu}{2}k\left(\nabla\left(\frac{w_1^{h,n+1} - w_1^{h,n}}{k}\right), \nabla v^h\right) \\
&+ \frac{\nu}{2}\delta^2 k\left(\nabla\left(\frac{\zeta_1^{h,n+1} - \zeta_1^{h,n}}{k}\right), \nabla v^h\right) + \frac{1}{2}(b^*(w_1^{h,n+1}, w_1^{h,n+1}, v^h) \\
&\quad - b^*(w_1^{h,n}, w_1^{h,n}, v^h)),
\end{aligned} \tag{4.4b}
$$

$$
(\nabla w_j^{h,n+1}, \nabla \xi^h) = (\zeta_j^{h,n+1}, \xi^h), \ j = 1, 2, \tag{4.4c}
$$

$$
(\nabla \cdot w_j^{h,n+1}, \chi^h) = 0, \ j = 1, 2. \tag{4.4d}
$$

The velocity field in turbulent flows is expected to be anything but smooth: with the flow's behaviour changing quickly and sporadically, the "true" velocity (exact solution of the NSE) is not expected to possess high regularity with respect to time or space. Yet, the numerical analysis of the ADMs produces theoretical results under the assumption of increased spatial regularity of the filtered velocity, compared to the "typical" regularity assumptions of the NSE velocity. This is common in hyperviscosity models, where higher spatial derivatives of the velocity are added to the NSE in an attempt to control the flow.

Increased regularity with respect to time is expected from the true solution, in order for the deferred correction to produce a higher accuracy solution. This is in line with the approach in the previous chapter, where a regularity result was needed for the time derivative of the error in the defect step solution (see also the regularity assumptions in Theorems 3 and 4 of Chapter 3).

Thus, the first question is to investigate the relevance of higher regularity assumptions on the true solution. It is also of interest to track the "asymptotic" nature of convergence, typical for general defect correction methods and, in particular, for the deferred correction. Does one need to refine the spatial mesh (and/or the time step) substantially, in order to see the increased convergence rates? Do the claimed convergence rates hold when the Reynolds number is increased? Before we get to testing the model (4.4) numerically, let us review the theoretical results from [57].

4.3 NUMERICAL STABILITY AND ERROR ESTIMATES

The following norm, induced by the deconvolution operator A, is used in the statements below.

$$\|\phi\|_A^2 = \|\phi\|^2 + \delta^2\|\nabla\phi\|^2.$$

Theorem 5 (Stability of the first approximation). *Let w_1^h satisfy (4.4a). Let $f \in L^2(0,T;H^{-1}(\Omega))$. Then, for $n = 0, \ldots, N-1$,*

$$\|w_1^{h,n+1}\|_A^2 + \nu k\Sigma_{i=0}^n\|\nabla w_1^{h,i+1}\|^2 + \nu\delta^2 k\Sigma_{i=0}^n\|\zeta_1^{h,i+1}\|^2$$
$$\leq \|w_1^{h,0}\|_A^2 + \frac{1}{\nu}k\Sigma_{i=0}^n\|f(t_{i+1})\|_{-1}^2.$$

The accuracy of the ADM is established with higher regularity assumptions (in space) on the filtered velocity field.

Theorem 6. *Let the time step satisfy*

$$k < \frac{\nu^3}{\max_{i=0,1,..,N}\|\nabla\bar{u}(t_i)\|^4}. \tag{4.5}$$

Also let $\bar{u} \in L^2(0,T;H^3(\Omega))$ and $\bar{u}_{tt} \in L^2(0,T;H^1(\Omega))$. Then, the error in the first approximation satisfies

$$\|\bar{u}(t_{n+1}) - w_1^{h,n+1}\|_A^2 + \nu k\sum_{i=0}^{n}\|\nabla(\bar{u}(t_{i+1}) - w_1^{h,i+1})\|^2 \tag{4.6}$$
$$+\nu\delta^2 k\sum_{i=0}^{n}\|\zeta(t_{i+1}) - \zeta_1^{h,i+1}\|^2 \leq C(\nu,\bar{u})(k^2 + \delta^4$$
$$+ k\sum_{i=0}^{n}((1+\delta^2h^{-2})\inf_{v\in V^h}\|\nabla(\bar{u}(t_i) - v^i)\|^2 + \inf_{q\in Q^h}\|p(t_i) - q^i\|^2)).$$

If the problem is linearized, the time step restriction can be substantially relaxed. This involves replacing one of the unknown velocities in the nonlinear term with a linear combination of the velocity values at previous time levels. While being computationally appealing, these approaches alter the mechanism of energy transport from large to small scales. In this book we will always keep the nonlinear term preserved.

4.3.1 Key Result to Ensure the Increased Accuracy

In line with the deferred correction approach, an estimate is needed for the time difference of the ADM velocity error, $\frac{e_1^{i+1}-e_1^i}{k} = \frac{(\bar{u}(t_{i+1})-w_1^{h,i+1})-(\bar{u}(t_i)-w_1^{h,i})}{k}$. The following theorem from [57] provides the corresponding result; notice the increase in the required temporal regularity of the true solution.

Theorem 7. *Let the conditions of Theorem 6 be satisfied. Also let $\bar{u}_t \in L^2(0,T;H^3(\Omega))$ and $\bar{u}_{ttt} \in L^2(0,T;H^1(\Omega))$. Then*

$$\|\frac{e_1^{n+1}-e_1^n}{k}\|_A^2 + \nu k \sum_{i=0}^{n} \|\nabla(\frac{e_1^{n+1}-e_1^n}{k})\|^2 \tag{4.7}$$

$$+\nu\delta^2 k \sum_{i=0}^{n} \|\frac{(\zeta(t_{i+1})-\zeta_1^{h,i+1})-(\zeta(t_i)-\zeta_1^{h,i})}{k}\|^2$$

$$\leq C(\nu,\bar{u})(k^2+\delta^4$$

$$+k\sum_{i=0}^{n}(\|\nabla\frac{(\bar{u}(t_{i+1})-\tilde{w}^{i+1})-(\bar{u}(t_i)-\tilde{w}^i)}{k}\|^2$$

$$+\inf_{q\in Q^h}\|\frac{(p(t_{i+1})-q^{i+1})-(p(t_i)-q^i)}{k}\|^2)),$$

where $\tilde{w}^{i+1}$ is the modified Stokes projection of $\bar{u}(t_{i+1})$.

Along with the accuracy result for the modified Stokes projection, [57], this proves that the time difference of the ADM error is first order accurate in time and second order accurate in δ. This suffices to prove the accuracy of the correction step - under the regularity assumptions of the theorems above.

Theorem 8. *The error in the second approximation satisfies*

$$\|\bar{u}(t_{n+1})-w_2^{h,n+1}\|_A^2 + \nu k \sum_{i=0}^{n} \|\nabla(\bar{u}(t_{i+1})-w_2^{h,i+1})\|^2 \tag{4.8}$$

$$+\nu\delta^2 k \sum_{i=0}^{n} \|\zeta(t_{i+1})-\zeta_2^{h,i+1}\|^2$$

$$\leq C(\nu,\bar{u})(k^4+\delta^4+\delta^2 k\sum_{i=0}^{n}\inf_{\chi\in X^h}\|\zeta(t_i)-\chi^i\|^2$$

$$+ k\sum_{i=0}^{n}(\inf_{v\in V^h}\|\nabla(\bar{u}(t_i)-v^i)\|^2+\inf_{q\in Q^h}\|p(t_i)-q^i\|^2)).$$

4.4 A NUMERICAL TEST

Numerical tests of models for turbulent fluid flows can be divided into three groups. In order to check the claimed convergence rates, one applies the model to a problem with a known true solution – and, usually, modifies some parameters (mesh and time step refinement) to check the dynamics of the results. Unfortunately, this means that the true solution is smooth, even though it might depend on a (possibly high) Reynolds number. The second group of tests pertains to the so-called transitional flows, where the Reynolds number is large enough for the flow to possess some interesting features and coherent structures, worth modeling. Finally, the last series of tests deals with fully developed, usually 3-D, turbulence. Experimental data is usually available for model's validation.

Herein, we merely scratch the surface of numerical testing - applying the model to a problem, [57], with the manufactured solution

$$u = \begin{pmatrix} 0.75 + 0.25\cos(2\pi(x-t))\sin(2\pi(y-t))e^{-8\pi^2 t/Re} \\ 0.75 - 0.25\sin(2\pi(x-t))\cos(2\pi(y-t))e^{-8\pi^2 t/Re} \end{pmatrix},$$

$$p = -\frac{1}{64}(\cos(4\pi(x-t))+\cos(4\pi(y-t)))e^{-16\pi^2 t/Re}.$$

The quantity of interest, approximated by the ADM, is $(\bar{u}, p)$. Thus, we filter the true solution and then compare it to the solutions (defect step solution w_1 and correction step solution w_2) obtained by our model. Taylor–Hood finite elements (piecewise quadratic polynomials for velocity and piecewise linear polynomials for pressure) are used for spatial discretization. The a priori error estimates predict the overall errors $\|\bar{u}-w_1\|_{L^2(0,T;L^2(\Omega))} = O(h^2+\delta^2+k)$ and $\|\bar{u}-w_2\|_{L^2(0,T;L^2(\Omega))} = O(h^2+\delta^2+k^2)$. Setting $\delta = k = h$ and refining the mesh diameter h, we expect to see first order accuracy for the defect step, and second order for the correction step solution. The Reynolds number is $Re = 10^5$.

There are two conclusions that could be drawn from Table 4.1. The convergence rates, at least in this relatively simple case, are consistent with the theoretical claims, and the deferred correction could thus be used to improve the temporal accuracy even in such a demanding and challenging application as modeling of fluid flows at high Reynolds numbers. Secondly, one could compare the absolute errors, obtained at the finest mesh for the defect step solution, and the penultimate error in the correction step approximations.

TABLE 4.1 Approximating the average solution

h	$\|\bar{u} - w_1\|_{L^2(0,T;L^2(\Omega))}$	*rate*	$\|\bar{u} - w_2\|_{L^2(0,T;L^2(\Omega))}$	*rate*
1/16	0.013047		0.0131178	
1/32	0.00584342	1.16	0.00417758	1.65
1/64	0.00309312	0.92	0.00114869	1.86
1/128	0.00167765	0.88	0.00029897	1.94

$Re = 10^5$, $T = 1$.

They are of similar magnitude (in fact, the correction step error at $h = 1/64$ is smaller than the defect step error at $h = 1/128$); therefore, one could use deferred correction to achieve the required accuracy faster – especially, if the defect and correction steps are computed in parallel (to be further discussed in the following chapters).

CHAPTER 5

Can Defect Correction Replace a Turbulence Model?

In the previous chapter we saw an attempt to "marry" general defect correction ideas and turbulence modeling. Therein, a deferred correction approach was used to reduce the error due to temporal discretization in the ADM turbulence model. This is one of the three components of the overall error: temporal discretization error, spatial discretization error (we will denote it the FE error, since all the spatial discretizations in this book stem from a finite element method) and the modeling error. The latter will become the focus of our attention in this chapter.

We start our discussions with equations (4.3) of the previous chapter. This formulation of the ADM seeks the approximations (w, q) to the pair $(\bar{u}, p)$ of filtered velocity and non-filtered pressure; the size of the system is substantially increased due to the added hyperviscosity term: the extra variable ζ is introduced in order to retain the C^0 finite element spaces.

5.1 CORRECTING THE ADM: FORMULATION AND EXPECTATIONS OF THE NEW METHOD

The author in [100] tried to use a defect correction method to reduce the modeling error of the ADM (4.3) and at the same time reduce the size of the system by replacing the hyperviscosity with the increased cell Reynolds number.

The key feature of the novel method, that separates it from, say, the defect correction idea in Chapter 2, is the treatment of the whole turbulence model as the "defect" that needs to be addressed in the correction step. In particular, since the ADM seeks to approximate the filtered velocity, the author started

DOI: 10.1201/9781003355977-5

by looking at the SFNSE - the spatially filtered formulation of the NSE. In it, there are four nonlinear terms

$$u \cdot \nabla u = \left(\bar{u} - \delta^2 \Delta \bar{u}\right) \cdot \nabla \left(\bar{u} - \delta^2 \Delta \bar{u}\right) \tag{5.1}$$
$$= \bar{u} \cdot \nabla \bar{u} - \delta^2 \bar{u} \cdot \nabla(\Delta \bar{u}) - \delta^2 \Delta \bar{u} \cdot \nabla \bar{u} + \delta^4 \Delta \bar{u} \cdot \nabla(\Delta \bar{u}).$$

Only the first of these terms has its counterpart in the left-hand side of (4.3). The three missing terms – the "defect" – are now restored in the correction step. In order to do so in the "defect correction way", these terms are functions of the defect solution, with its laplacian computed in the postprocessing of the defect step.

This resulted in the following defect correction method (DCM) to approximate $(\bar{u}, p)$. Its semi-discrete (continuous in time, discretized in space) formulation reads: find successive approximations $(w_1^h, q_1^h), (w_2^h, q_2^h) \in (X^h, Q^h)$ satisfying for $\forall (v^h, \xi^h) \in (X^h, Q^h)$

$$\begin{aligned}
(w_{1,t}^h, v^h) + \delta^2(\nabla w_{1,t}^h, \nabla v^h) + (\nu + \delta)(\nabla w_1^h, \nabla v^h) + b^*(w_1^h, w_1^h; v^h)& \\
-(q_1^h, \nabla \cdot v^h) = (f, v^h),& \\
(w_{2,t}^h, v^h) + \delta^2(\nabla w_{2,t}^h, \nabla v^h) + (\nu + \delta)(\nabla w_2^h, \nabla v^h) + b^*(w_2^h, w_2^h; v^h)& \\
-(q_2^h, \nabla \cdot v^h) = (f, v^h) + \delta(\nabla w_1^h, \nabla v^h) + \nu\delta^2(\nabla \zeta_1^h, \nabla v^h)& \\
+\delta^2 b^*(\zeta_1^h, w_1^h; v^h) + \delta^2 b^*(w_1^h, \zeta_1^h; v^h) - \delta^4 b^*(\zeta_1^h, \zeta_1^h; v^h),& \\
(\zeta_i^h, v^h) + (\nabla w_i^h, \nabla v^h) = 0, i = 1, 2,& \\
(\nabla \cdot w_i^h, \xi^h) = 0, i = 1, 2.&
\end{aligned} \tag{5.2}$$

Before we present the stability and accuracy results from [100], let's assess the terms in (5.2). The hyperviscosity terms of the ADM were replaced with an extra viscosity term of the size $O(\delta)$. Thus, we expect the defect step to be first order accurate in the filtering width (modeling error); the right-hand side of the correction step is adjusted to "correct" for this non-physical viscosity, so we hope that the solution of the correction step is second order accurate in δ.

Note also that this formulation could benefit from the deferred correction approach of Chapter 4. A lower order, stable and computationally appealing method could be used for the time discretization of the defect step in (5.2); deferred correction would fit naturally into the correction step of (5.2), reducing the temporal discretization error by adding more (known) terms to the right-hand side. In this chapter, we only focus on the modeling error, as did the author in [100], and we use the Backward Euler time discretization without the deferred correction. Combining (5.2) with the method of Chapter 4 could lead an interested reader to a novel, efficient and accurate regularization technique.

5.2 NUMERICAL ANALYSIS

The fully discretized version of (5.2), that utilizes the Backward Euler method for time discretization, seeks $(w_1^{h,n+1}, q_1^{h,n+1})$, $(w_2^{h,n+1}, q_2^{h,n+1})$ such that the following equations are satisfied $\forall(v^h, \xi^h) \in (X^h, Q^h)$.

$$
\begin{aligned}
&(\frac{w_1^{h,n+1} - w_1^{h,n}}{k}, v^h) + \delta^2(\frac{\nabla w_1^{h,n+1} - \nabla w_1^{h,n}}{k}, \nabla v^h) \\
&+(\nu+\delta)(\nabla w_1^{h,n+1}, \nabla v^h) + b^*(w_1^{h,n+1}, w_1^{h,n+1}; v^h) \\
&-(q_1^{h,n+1}, \nabla \cdot v^h) = (f(t_{n+1}), v^h), \\
&(\frac{w_2^{h,n+1} - w_2^{h,n}}{k}, v^h) + \delta^2(\frac{\nabla w_2^{h,n+1} - \nabla w_2^{h,n}}{k}, \nabla v^h) \\
&+(\nu+\delta)(\nabla w_2^{h,n+1}, \nabla v^h) + b^*(w_2^{h,n+1}, w_2^{h,n+1}; v^h) \\
&-(q_2^{h,n+1}, \nabla \cdot v^h) = (f(t_{n+1}), v^h) + \delta(\nabla w_1^{h,n+1}, \nabla v^h) \\
&+\nu\delta^2(\nabla \varsigma_1^{h,n+1}, \nabla v^h) + \delta^2 b^*(\varsigma_1^{h,n+1}, w_1^{h,n+1}; v^h) \\
&+\delta^2 b^*(w_1^{h,n+1}, \varsigma_1^{h,n+1}; v^h) - \delta^4 b^*(\varsigma_1^{h,n+1}, \varsigma_1^{h,n+1}; v^h), i = 1, 2, ..., \\
&(\varsigma_i^{h,n+1}, v^h) + (\nabla w_i^{h,n+1}, \nabla v^h) = 0, i = 1, 2, \\
&(\nabla \cdot w_i^{h,n+1}, \xi^h) = 0, i = 1, 2.
\end{aligned} \tag{5.3}
$$

The numerical analysis, presented in [100], is given in the five theorems below. It starts with the stability and accuracy of the defect step approximation.

Theorem 9 (Stability of the Defect Step approximation). *Let w_1^h satisfy the first equation of (5.3). Let $f \in L^2(0,T;H^{-1}(\Omega))$. Then for $n = 0, ..., N-1$*

$$
\|w_1^{h,n+1}\|^2 + \delta^2\|\nabla w_1^{h,n+1}\|^2 + (\delta+\nu)k\Sigma_{i=1}^{n+1}\|\nabla w_1^{h,i}\|^2 \le \|w_1^{h,0}\|^2 + \delta^2\|\nabla w_1^{h,0}\|^2 + \frac{1}{\delta+\nu}k\Sigma_{i=1}^{n+1}\|f(t_i)\|_{-1}^2.
$$

Also, if $f \in L^2(0,T;L^2(\Omega))$ and the time constraint T is finite, then there exists a constant $C = C(T)$ such that

$$
\begin{aligned}
&\|w_1^{h,n+1}\|^2 + \delta^2\|\nabla w_1^{h,n+1}\|^2 + (\delta+\nu)k\Sigma_{i=1}^{n+1}\|\nabla w_1^{h,i}\|^2 \\
&\le C(\|w_1^{h,0}\|^2 + \delta^2\|\nabla w_1^{h,0}\|^2 + k\Sigma_{i=1}^{n+1}\|f(t_i)\|^2).
\end{aligned} \tag{5.4}
$$

Theorem 10 (Accuracy of the Defect Step approximation). *Let the true solution satisfy*

$$
\begin{aligned}
&\bar{u} \in L^2(0,T;H^{m+1}(\Omega)) \bigcap L^\infty(0,T;L^\infty(\Omega)), \nabla\bar{u} \in L^\infty(0,T;L^\infty(\Omega)), \\
&\bar{u}_t \in L^2(0,T;H^{m+1}(\Omega)), \bar{u}_{tt} \in L^2(0,T;L^2(\Omega)), p \in L^2(0,T;H^m(\Omega)).
\end{aligned}
$$

Then there exists a constant $C = C(\Omega, T, u, p, f)$, such that, under the time step restriction $k \leq C\left(\delta + Re^{-1}\right)$, the defect step approximation w_1^h of (5.3) satisfies

$$\max_{1 \leq i \leq N} \left(\|\bar{u}(t_i) - w_1^{h,i}\|^2 + \delta^2 \|\nabla(\bar{u}(t_i) - w_1^{h,i})\|^2 \right)$$
$$+k \sum_{i=1}^{n+1} (\delta + \nu) \|\nabla(\bar{u}(t_i) - w_1^{h,i})\|^2 \leq C(h^{2m} + \delta^2 + k^2).$$

Next we establish the accuracy of the gradient error - the improvements in the modeling error may be achieved only if we can show that the gradient of the true velocity is approximated well enough by the gradient of the defect step solution.

Theorem 11 (Gradient error). *Let the conditions of Theorem 10 be satisfied. Assume also the increased regularity of the true solution,*

$$\Delta \bar{u} \in L^\infty(0, T; L^\infty(\Omega)), \nabla^3 \bar{u} \in L^\infty(0, T; L^\infty(\Omega)).$$

Then there exists a constant $C = C(\Omega, T, u, p, f)$, such that, under the (tighter) time step restriction $k \leq Ch\left(\delta + Re^{-1}\right)$, the defect step approximation w_1^h of (5.3) satisfies

$$\max_{1 \leq i \leq N} \left(\|\nabla(\bar{u}(t_i) - w_1^{h,i})\|^2 + \delta^2 \|\Delta \bar{u}(t_i) - \zeta_1^{h,i}\|^2 \right)$$
$$+k \sum_{i=1}^{n+1} (\delta + \nu) \|\Delta \bar{u}(t_i) - \zeta_1^{h,i}\|^2 \leq C(h^{2(m-1)} + \delta^2 + k^2).$$

It should not be surprising that these "intermediate" theorems establishing the accuracy of higher spatial (as in this chapter) or temporal (previous chapter) derivatives of the defect step solution, are the most subtle point in every proof mentioned in this book. These theorems impose tighter regularity requirements on the problem data, and may also ask for a smaller time step or smaller filtering width. Once the accuracy of higher derivatives is established, proving the accuracy of the correction step approximation is almost straightforward.

Theorem 12 (Stability of the Correction Step approximation). *Let w_2^h satisfy (5.3). Then, under the conditions of Theorem 11, $\forall n = 0, ..., N-1$:*

$$\|w_2^{h,n+1}\|^2 + \delta^2 \|\nabla w_2^{h,n+1}\|^2 + k\Sigma_{i=1}^{n+1} (\delta + \nu) \|\nabla w_2^{h,i}\|^2$$
$$\leq \|w_2^{h,0}\|^2 + \delta^2 \|\nabla w_2^{h,0}\|^2 + \frac{1}{\delta + \nu} k \Sigma_{l=1}^{n+1} \|f(t_l)\|_{-1}^2 + C(\Omega, \bar{u}).$$

Theorem 13 (Accuracy of the second approximation)**.** *Let w_2^h satisfy (5.3). Then, under the conditions of Theorem* 11*, there exists a constant* $C = C(\Omega, T, u, p, f)$*, such that*

$$\max_{1\leq i\leq N}\left(\|\bar{u}(t_i) - w_2^{h,i}\|^2 + \delta^2\|\nabla(\bar{u}(t_i) - w_2^{h,i})\|^2\right)$$

$$+k\sum_{i=1}^{n+1}(\delta+\nu)\|\nabla(\bar{u}(t_i) - w_2^{h,i})\|^2 \leq C(h^{2m} + \delta^4 + k^2).$$

5.3 NUMERICAL TESTS OF THE CONVERGENCE RATES

The method, presented in [100] and in this chapter, should be viewed as a numerical regularization of flows at high Reynolds numbers - at least until it is tested on some of the "canonical" test problems for incompressible turbulence (e.g., a turbulent channel flow problem). The first thing to test is the method's efficiency: it should outperform its most natural competitors in terms of "accuracy per computational cost". That is, the method should require less computational resources to produce the solution of desired quality (for problems with known true solution this would mean reaching the prescribed tolerance) – or produce a higher quality solution when compared to other methods' solutions obtained at the prescribed computational cost (on the same mesh, with the same time step, in the same CPU time, etc.).

The most natural competitor for the defect correction method (5.2) is the ADM. In [100] the two approaches were compared on a simple 2-D "traveling wave" problem in the region $\Omega = [0.5, 1]^2$, with a known true solution. The parameters were chosen as follows: $\nu = 10^{-4}$, $T = 1$, $\Delta t = \delta = h$. Table 5.1 below shows the L^2 norms of the error of the ADM solution (4.3) and the correction step solution of (5.3). The author of [100] used the higher accurate Crank–Nicolson method for time discretization of (4.3), even though the chosen time discretization method in (5.3) was Backward Euler. Thus, the second order accurate (in time and in filtering width) ADM looks to be a preferable method for reducing both the modeling error and the temporal discretization error - but Table 5.1 paints a different picture.

The Taylor-Hood finite elements were used for spatial discretization. The novel method (5.3) of this chapter seems to produce a smaller error than the ADM at every mesh where the ADM solution is computed. Eventually the direct solver UMFPACK fails to compute the ADM solution when the mesh is too fine; the defect correction method is still within range for the UMFPACK solver, because the size of the system is significantly smaller for (5.2) than it is for (4.3). This computational attractiveness is also evident when we compare the CPU time needed for the two methods. Note that the defect and correction steps were taken sequentially, and still the CPU time needed for the method (5.3) is about 80% of the time that is needed for the ADM. With the defect and correction step solvers running in parallel on two cores (with the correction step lagging in time, so that $w_1^{h,n+1}$ and $w_2^{h,n}$ are

TABLE 5.1 Averaged true solution $\bar{u}$ vs. the Correction Step solution w_2^h and the ADM solution w^h_{ADM}

$1/h$	$\|w^h_{ADM} - \bar{u}\|_{L^2(0,T;L^2(\Omega))}$	ADM CPU time	$\|w_2^h - \bar{u}\|_{L^2(0,T;L^2(\Omega))}$	DC CPU time
8	0.0210767	4.4	0.0173342	3.9
16	0.0122257	34	0.00873841	28
32	0.00522025	252	0.0033991	196
64	0.00156767	2077	0.00122641	1642
128	UMFPACK failed	–	0.000441361	13292

computed in parallel) it is reasonable to expect that the novel method would run in less than half of the time needed for the ADM.

Of course, more tests are required to highlight both the pros and the cons of this new approach. These could include more of 2-D and 3-D qualitative tests, and, eventually, the benchmark problems for incompressible turbulence.

CHAPTER 6

Deferred Correction for a Fluid-Fluid Interaction Problem

In this chapter we turn our attention to a coupled Navier–Stokes system – a fluid–fluid interaction problem, [105]. Deferred correction methods promise to produce solutions of higher accuracy, but this could also be achieved by Runge–Kutta methods or multistep methods of the corresponding order. The main advantage of deferred correction is the fact that this increased accuracy is achieved despite using low order methods with the desired stability properties. In particular, this makes the deferred correction a very attractive approach for increasing the accuracy of partitioning methods. Breaking a large coupled system into a pair of subsystems is generally a very hard task; partitioning methods are usually developed that are stable, yet only low order accurate.

6.1 FLUID–FLUID INTERACTION – A SIMPLIFIED MODEL FOR ATMOSPHERE–OCEAN COUPLING

In this chapter we will discuss two Navier–Stokes equations, that govern the flow of two fluids in the domains that share a joint interface. Consider the two-dimensional domain Ω that consists of two subdomains Ω_1 and Ω_2, coupled across an interface I, as shown in Figure 6.1. Fluid–fluid and fluid–structure interaction problems in a similar setting can be found in [111, 113].

The problem reads: *given* $\nu_i > 0$, $f_i : [0,T] \to H^1(\Omega_i)^d, u_i(0) \in H^1(\Omega_i)^d$ *and* $\kappa \in \mathbb{R}$, *find (for* $i = 1,2$*)* $u_i : \Omega_i \times [0,T] \to \mathbb{R}^d$ *and* $p_i : \Omega_i \times [0,T] \to \mathbb{R}$ *satisfying (for* $0 < t \leq T$*)*

$$u_{i,t} = \nu_i \Delta u_i - u_i \cdot \nabla u_i - \nabla p_i + f_i, \quad \text{in } \Omega_i, \tag{6.1}$$
$$-\nu_i \hat{n}_i \cdot \nabla u_i \cdot \tau = \kappa |u_i - u_j|(u_i - u_j) \cdot \tau, \quad \text{on } I, \ i,j = 1,2, \ i \neq j, \tag{6.2}$$

DOI: 10.1201/9781003355977-6

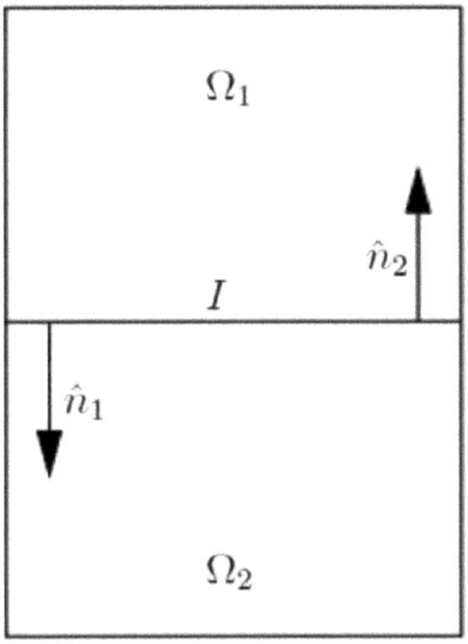

Figure 6.1 Example subdomains, coupled across an interface I.

$$u_i \cdot \hat{n}_i = 0 \quad \text{on } I, i = 1,2 \tag{6.3}$$
$$\nabla \cdot u_i = 0, \quad \text{in } \Omega_i \tag{6.4}$$
$$u_i(x,0) = u_i^0(x), \quad \text{in } \Omega_i, \tag{6.5}$$
$$u_i = 0, \quad \text{on } \Gamma_i = \partial\Omega_i \setminus I. \tag{6.6}$$

The vectors $\hat{n}_i$ are the unit normals on $\partial\Omega_i$, and τ is any vector such that $\tau \cdot \hat{n}_i = 0$; $\kappa > 0$ is a friction parameter.

The coupling condition (6.2) was introduced and studied in [130]; it represents the flux of momentum across a boundary layer region near the fluid interface. Along with the condition (6.3) it basically says, referring to the atmosphere–ocean coupling: no water in the air, no air in the water, the energies of the subsystems can be passed across the interface, but the overall energy of the coupled system is conserved.

Reasons for our interest in the fluid–fluid interaction problem (6.1)–(6.6) are detailed in [105]. This system of equations is not sophisticated enough to model real-life phenomena related to the interaction of air and sea, but it does possess some key features of the climate models currently in use, including the nonlinear coupling condition at the joint interface. Given the possible turbulent nature of one or both flows, this presents a unique set of challenges.

6.1.1 Importance of a Partitioning Method

First and foremost, one would like to decouple the system in order to use the pre-existing codes ("legacy codes") for resolving the atmospheric and oceanic parts of the domain. Due to the complex geometries of the shoreline and the possible need for turbulence modeling, such legacy codes can prove to be invaluable. Another reason for decoupling is obvious: solving one large $2N$-by-$2N$ system is much more challenging computationally, than solving two N-by-N subsystems.

A partitioning method for (6.1)–(6.6) was presented in [118]. The authors called it the Geometric Averaging method (GA), due to its treatment of the interface condition. They proved the unconditional stability of this method – and to date this remains (to the best of our knowledge) the best decoupling method for (6.1)–(6.6) in terms of stability. However, GA is only first order accurate in time; deferred correction to the rescue. The authors of [105] recognized this as an opportunity for deferred correction to shine: the "usual suspects" from the Runge–Kutta and multistep families would not decouple (6.1)–(6.6), whereas a deferred correction might be built on the stable GA method and improve its accuracy. Additionally, the correction step could be used to improve the accuracy of a numerical regularization technique, aimed at controlling the flow in a turbulent setting.

6.2 DEFECT–DEFERRED CORRECTION METHOD

The authors of [105] looked at a very simple yet fairly efficient artificial viscosity approximation, discussed in Chapter 2. It is not sophisticated enough to be called a turbulence model, but this regularization technique allows to obtain a solution at a high Reynolds number – even though the accuracy of such a solution needs improvement. Thus, a method proposed in [105] had two key ingredients. Based on the GA partitioning approach of [118], it proposed to take one correction step to reduce not one, but two "defects": improve the temporal accuracy of the GA via the deferred correction, and at the same time reduce the modeling error via defect correction.

Therein lie the issues that the authors of [105] had to overcome. The numerical analysis for the defect-deferred correction procedure is tedious and requires the accuracy estimates of the temporal derivatives of the velocities. The extra nonlinearity in the interface condition (6.2) increases the complexity of the problem. On top of these issues, few benchmark problems existed in the literature for this fluid–fluid interaction, so the authors of [105] had to come up with new settings for numerical tests.

Introduce the function spaces for the subdomains.

For $i = 1, 2$, let

$$X_i := \{v_i \in H^1(\Omega_i)^d : v_i = 0\, on\, \Gamma_i, i = 1, 2, v_i \cdot \hat{n}_i = 0\, on\, I\}$$

be velocity spaces, with associated pressure spaces

$$Q_i = \{q_i \in L^2(\Omega_i)^d : \int_{\Omega_i} q_i d\Omega_i = 0\}.$$

The weak formulation of the problem (6.1) - (6.6) aims to find (for $i, j =$

$1, 2, i \neq j$) $u_i : [0, T] \to X_i$ and $p_i : [0, T] \to Q_i$ satisfying

$$
\begin{aligned}
(u_{i,t}, v_i)_{\Omega_i} + \nu_i(\nabla u_i, \nabla v_i)_{\Omega_i} + (u_i \cdot \nabla u_i, v_i)_{\Omega_i} - (p_i, \nabla v_i) \\
+ \int_I \kappa(u_i - u_j)|u_i - u_j| v_i ds = (f_i, v_i)_{\Omega_i}, \ \forall\, v_i \in X_i, \\
(\nabla \cdot u_i, q_i) = 0, \ \forall\, q_i \in Q_i.
\end{aligned}
\tag{6.7}
$$

For functions $u, v, w \in X_i$, $i = 1, 2$ define the expilicitly skew-symmetrized nonlinear form on Ω_i by

$$
b_i(u; v, w) = \frac{1}{2}(u \cdot \nabla v, w)_{\Omega_i} - \frac{1}{2}(u \cdot \nabla w, v)_{\Omega_i} \tag{6.8}
$$

We are now ready to define the Geometric Averaging (GA) method of [118], which was shown by the authors to be unconditionally stable and first order accurate.

Let $\Delta t > 0$, $f_i \in L^2(\Omega_i)$. For each $M \in \mathbb{N}, M \leq \frac{T}{\Delta t}$, given $u_i^n \in X_{i,h}$ and $p_i^n \in Q_{i,h}$, $n = 0, 1, 2, \cdots, M - 1$, solve on each subdomain (for $i, j = 1, 2, i \neq j$) to find $u_i^{n+1} \in X_{i,h}$ satisfying

$$
\begin{aligned}
\left(\frac{u_i^{n+1} - u_i^n}{\Delta t}, v_i \right) + \nu_i(\nabla u_i^{n+1}, \nabla v_i) \\
+ \kappa \int_I u_i^{n+1} |u_i^n - u_j^n| v_i \, ds - \kappa \int_I u_j^n |u_i^n - u_j^n|^{1/2} |u_i^{n-1} - u_j^{n-1}|^{1/2} v_i \, ds \\
+ b_i(u_i^{n+1}; u_i^{n+1}, v_i) - (p_i^{n+1}, \nabla \cdot v_i) = (f_i(t^{n+1}), v_i), \ \forall v_i \in X_{i,h}\,. \\
(\nabla \cdot u_i^{n+1}, q_i) = 0, \ \forall\, q_i \in Q_{i,h}.
\end{aligned}
\tag{6.9}
$$

6.2.1 Formulation of the Method

The defect-deferred correction method of [105], that aims to improve the temporal accuracy of the GA and adds (and corrects for) an artificial viscosity to tackle flows at high Reynolds numbers, is presented via the following procedure.

Algorithm 6.2.1 (Two Step DDC). *Let* $\Delta t > 0$, $M = \frac{T}{\Delta t}$, $f_i \in L^2(\Omega_i)$. *Given* $\hat{u}_i^n$, *find* $\hat{u}_i^{n+1} \in X_i^h$, $i, j = 1, 2, i \neq j$, $n = 0, 1, 2, \cdots, M - 1$, *satisfying*

$$
\begin{aligned}
\left(\frac{\hat{u}_i^{n+1} - \hat{u}_i^n}{\Delta t}, v_i \right) + (\nu_i + H_i) \left(\nabla \hat{u}_i^{n+1}, \nabla v_i \right) \\
+ \kappa \int_I |[\hat{u}^n]| \hat{u}_i^{n+1} v_i ds - \kappa \int_I \hat{u}_j^n |[\hat{u}^n]|^{1/2} |[\hat{u}^{n-1}]|^{1/2} v_i ds \\
- (\hat{p}_i^{n+1}, \nabla \cdot v_i) + b_i \left(\hat{u}_i^{n+1}; \hat{u}_i^{n+1}, v_i \right) = \left(f_i^{n+1}, v_i \right), \ \forall v_i \in X_{i,h}
\end{aligned}
\tag{6.10}
$$

Then, given $\hat{u}_i^{n+1}$ *and* $\tilde{u}_i^n$, *find* $\tilde{u}_i^{n+1} \in X_i^h$ *satisfying*

$$\left(\frac{\tilde{u}_i^{n+1}-\tilde{u}_i^n}{\Delta t}, v_i\right) + (\nu_i + H_i)\left(\nabla \tilde{u}_i^{n+1}, \nabla v_i\right)$$
$$+\kappa \int_I |[\tilde{u}^n]| \tilde{u}_i^{n+1} v_i ds - \kappa \int_I \tilde{u}_j^n |[\tilde{u}^n]|^{1/2} |[\tilde{u}^{n-1}]|^{1/2} v_i ds - (\tilde{p}_i^{n+1}, \nabla \cdot v_i)$$
$$+b_i\left(\tilde{u}_i^{n+1}; \tilde{u}_i^{n+1}, v_i\right) = \left(\frac{f_i^{n+1}+f_i^n}{2}, v_i\right) + \frac{\Delta t(\nu_i + H_i)}{2}\left(\nabla(\frac{\hat{u}_i^{n+1}-\hat{u}_i^n}{\Delta t}), \nabla v_i\right)$$
$$+\frac{\kappa}{2}\Delta t \int_I |[\hat{u}^n]| (\frac{\hat{u}_i^{n+1}-\hat{u}_i^n}{\Delta t}) v_i ds - \frac{\kappa}{2}\Delta t \int_I \hat{u}_i^{n+1} (\frac{|[\hat{u}^{n+1}]| - |[\hat{u}^n]|}{\Delta t}) v_i ds$$
$$+ H_i\left(\nabla(\frac{\hat{u}_i^{n+1}+\hat{u}_i^n}{2}), \nabla v_i\right) - \kappa \int_I \hat{u}_j^n |[\hat{u}^n]|^{1/2} |[\hat{u}^{n-1}]|^{1/2} v_i ds$$
$$+\frac{\kappa}{2}\int_I |[\hat{u}^{n+1}]| \hat{u}_j^{n+1} v_i ds + \frac{\kappa}{2}\int_I |[\hat{u}^n]| \hat{u}_j^n v_i ds + \frac{1}{2} b_i(\hat{u}_i^{n+1}; \hat{u}_i^{n+1}, v_i)$$
$$-\frac{1}{2} b_i(\hat{u}_i^n; \hat{u}_i^n, v_i) - \left(\frac{\hat{p}_i^{n+1}-\hat{p}_i^n}{2}, \nabla \cdot v_i\right), \quad \forall v_i \in X_{i,h}. \quad (6.11)$$

6.3 NUMERICAL ANALYSIS

In order to proceed with the stability and accuracy results of [105], we introduce the notation for the global (monolithic) variables: let $\mathbf{u} = (u_1, u_2)$ and define $[\mathbf{u}] \equiv u_1 - u_2$. The following definition and assumptions were used by the authors of [105] in order to complete the proofs.

Definition 1. *Let* $i = 1, 2$.

$$C_u := \|\mathbf{u}(x,t)\|_{L^\infty(0,T;L^\infty(\Omega))},$$
$$C_{u_{i,t}} := \|\mathbf{u}_{i,t}(x,t)\|_{L^\infty(0,T;L^\infty(\Omega))},$$
$$C_{\nabla u_{i,t}} := \|\nabla \mathbf{u}_{i,t}(x,t)\|_{L^\infty(0,T;L^\infty(\Omega))}.$$

Assumption 1. $\exists \alpha > 0$, *such that* $\alpha \leq |[\mathbf{u}(\vec{x}, t)]|$, $\forall \vec{x} \in I$, $\forall t \in (0, T]$.

Assumption 2. *Let the true solution* u *satisfy*

$$\left|\frac{\partial}{\partial t}(|\mathbf{u}_i(t)|)\right| \leq C\Delta t^{1/4}, \text{for } i = 1, 2, 0 < t \leq \Delta t, \forall \vec{x} \in I. \quad (6.12)$$

6.3.1 Defect Step Results

We begin with the unconditional stability of the defect step approximation.

Theorem 1. *(Stability of Defect approximation) Let* $\hat{u}_i^j \in X_{i,h}$ *satisfy (6.10) for each* $j \in \{0, 1, 2, \ldots, \frac{T}{\Delta t} - 1\}$, $i = 1, 2$. *Then* $\exists C > 0$ *independent of* h, Δt *such that* $\hat{\mathbf{u}}^{n+1}$ *satisfies:*

$$\|\hat{\mathbf{u}}^{n+1}\|^2 + \sum_{j=1}^{n} \|\hat{\mathbf{u}}^{j+1} - \hat{\mathbf{u}}^j\|^2$$

$$+\Delta t\sum_{j=1}^{n}\left[(\nu_1+H_1)\|\nabla u_1^{j+1}\|^2_{\Omega_1}+(\nu_2+H_2)\|\nabla u_2^{j+1}\|^2_{\Omega_2}\right]$$

$$+\kappa\Delta t\int_I|[\hat{\mathbf{u}}^n]|(|u_1^{n+1}|^2+|u_2^{n+1}|^2)ds+\kappa\Delta t\sum_{j=1}^{n}\int_I|u_1^{j+1}|[\hat{\mathbf{u}}^j]|^{1/2}-u_2^j|[\hat{\mathbf{u}}^{j-1}]|^{1/2}|^2ds$$

$$+\kappa\Delta t\sum_{j=1}^{n}\int_I|u_2^{j+1}|[\hat{\mathbf{u}}^j]|^{1/2}-u_1^j|\hat{\mathbf{u}}^{j-1}|^{1/2}|^2ds\le\|\mathbf{u}^1\|^2+\kappa\Delta t\int_I|[\mathbf{u}^0]|(|u_1^1|^2+|u_2^1|^2)ds$$

$$+\sum_{j=1}^{n}\left[\frac{\Delta t}{\nu_1+H_1}\|f_1^{j+1}\|^2_{-1}+\frac{\Delta t}{\nu_2+H_2}\|f_2^{j+1}\|^2_{-1}\right].\quad(6.13)$$

The following accuracy result (derived in [105], but very similar to the accuracy estimate in [118] due to the similarity of GA and the defect step of Algorithm 6.2.1) sets the tone for the remainder of this Chapter. As is always the case for defect and deferred correction methods, the accuracy (and sometimes the stability) of the correction step solution hinges upon the accuracy result obtained for the defect step solution. The GA method is, as we have previously mentioned, unconditionally stable, but there are restrictions on the time step for this method to be first order accurate. These restrictions will transfer onto the correction step solution of Algorithm 6.2.1.

Theorem 2. *(Accuracy of Defect Solution) Let $\hat{u}_i^k \in X_{i,h}$ satisfy (6.10) for each $k \in 2,\dots,n\le N-1$. Let $\tilde{\nu}=max\{\nu_1^{-1},\nu_2^{-1}\}$, $\hat{\nu}=max\{\nu_1,\nu_2\}$, and let $D^{n+1}=\tilde{\nu}^3(1+\kappa^4E^{n+1}+\|\nabla\mathbf{u}^{n+1}\|^4)$, where $E^{n+1}=max_{j=0,1,\dots,n+1}\{\|\mathbf{u}^j\|_I^4\}$.Assume $\Delta t\le\frac{1}{D^{n+1}}$, and that $(\mathbf{u},\mathbf{p})$ is a strong solution of the coupled NSE system (6.1)-(6.6) with $\mathbf{u}_t\in L^2(0,T;X)$ and $\mathbf{u}_{tt}\in L^2(0,T;L^2(\Omega))$. Then the solution $\hat{\mathbf{u}}^{n+1}$ of (6.10) satisfies:*

$$\|\mathbf{u}^{n+1}-\hat{\mathbf{u}}^{n+1}\|^2+\frac{\Delta t}{2}(\nu_1+H_1)\sum_{k=1}^{n}\|\nabla(u_1^{k+1}-\hat{u}_1^{k+1})\|^2_{\Omega_1}$$

$$+\frac{\Delta t}{2}(\nu_2+H_2)\sum_{k=1}^{n}\|\nabla(u_2^{k+1}-\hat{u}_2^{k+1})\|^2_{\Omega_2}$$

$$\le Cexp\left\{\Delta t\sum_{n=1}^{j}\frac{D^{n+1}}{1-\Delta tD^{n+1}}\right\}\Bigg[\|\mathbf{u}^1-\hat{\mathbf{u}}^1\|^2$$

$$+\inf_{\mathbf{v}^1\in V_h}\|\mathbf{u}^1-\mathbf{v}^1\|^2+\Delta t(\nu_1+H_1)(\|\nabla(u_1^1-\hat{u}_1^1)\|^2_{\Omega_1}+\frac{1}{2}\|\nabla(u_1^0-\hat{u}_1^0)\|^2_{\Omega_1}$$

$$+\inf_{v_1^1\in V_{1,h}}\|\nabla(u_1^1-v_1^1)\|^2_{\Omega_1}+\frac{1}{2}\inf_{v_1^0\in V_{1,h}}\|\nabla(u_1^0-v_1^0)\|^2_{\Omega_1})+\Delta t(\nu_2+H_2)(\|\nabla(u_2^1-\hat{u}_2^1)\|^2_{\Omega_2}$$

$$+\frac{1}{2}\|\nabla(u_2^0-\hat{u}_2^0)\|^2_{\Omega_2}+\inf_{v_2^1\in V_{2,h}}\|\nabla(u_2^1-v_2^1)\|^2_{\Omega_2}+\frac{1}{2}\inf_{v_2^0\in V_{2,h}}\|\nabla(u_2^0-v_2^0)\|^2_{\Omega_2})$$

$$+\Delta t^2\|\mathbf{u}_{tt}\|^2_{L^2(0,T;L^2(\Omega))}+\inf_{\mathbf{v}\in V_h}\|(\mathbf{u}-\mathbf{v})_t\|^2_{L^2(0,T;L^2(\Omega))}$$

$$+\inf_{\mathbf{q}\in Q_h}\|\mathbf{p}-\mathbf{q}\|^2_{L^2(0,T;L^2(\Omega))}+\kappa^2\Delta t^2\|\mathbf{u}_t\|^2_{L^2(0,T;X)}$$

$$+T\max_{k=2,\cdots,n+1}\left(\inf_{\mathbf{v}^k\in V_h}\|\nabla(\mathbf{u}^k-\mathbf{v}^k)\|^2\right)+\frac{H_1^2}{\nu_1+H_1}C^2_{\nabla u_1}+\frac{H_2^2}{\nu_2+H_2}C^2_{\nabla u_2}\Bigg],\tag{6.14}$$

where C has the following dependence on κ , ν_1 and ν_2: $C = O(max\{\tilde{\nu},\hat{\nu},(1+\Delta t\kappa^4)\tilde{\nu}^3,\kappa^2\})$.

Assuming enough regularity of the problem data, the accuracy of the defect step solution is given by

Corollary 1. *Let the problem data be smooth enough; let the discrete velocity-pressure spaces consist of continuous piecewise polynomials of degrees m and $m-1$, respectively ($m \geq 2$). Then there exists a constant C independent of $h, H, \Delta t$, s.t.*

$$\|\mathbf{u}^{n+1}-\hat{\mathbf{u}}^{n+1}\|^2+\frac{\Delta t}{2}(\nu_1+H_1)\sum_{k=1}^{n}\|\nabla(u_1^{k+1}-\hat{u}_1^{k+1})\|^2_{\Omega_1}$$
$$+\frac{\Delta t}{2}(\nu_2+H_2)\sum_{k=1}^{n}\|\nabla(u_2^{k+1}-\hat{u}_2^{k+1})\|^2_{\Omega_2}$$
$$\leq C(h^{2m}+\Delta t^2+H_1^2+H_2^2).\tag{6.15}$$

6.3.2 Accuracy of the Derivative of the Defect Step Error

The main result of this Chapter, upon which hinges the success of the deferred correction algorithm, is given in the theorem below. Note that the assumptions on the size of the jump $|[\mathbf{u}]|$ of the true solution and the bound on the time derivative along the first time step were needed for this particular proof.

Consider $e_i^j = u_i^j - \hat{u}_i^j$, $i = 1, 2$, $j = 0, 1, 2, \ldots, M$.

Theorem 3. *(Accuracy of Time Derivative of the Error in the Defect Step) Let $u_i(\Delta t) \in H^2(\Omega_i)$, $\Delta\mathbf{u} \in L^2(0,T;L^2(\Omega))$ and $\mathbf{u}_{tt}, \mathbf{u}_t, \mathbf{u} \in L^2(0,T;L^2(\Omega))$. Let $min(h,\Delta t) < C(\frac{\nu_i+h_i}{\kappa})$.*

Let also $max(h,\Delta t,H_1,H_2) \leq \frac{\alpha}{4\sqrt{C_f}}$, where α is the constant introduced in Assumption 1, and C_f is the constant from (6.15).

Then $\exists C > 0$ independent of $h, H_i, \Delta t$ such that for any $n \in \{0,1,2,\ldots,M-1 = \frac{T}{\Delta t}-1\}$, the discrete time derivative of the error $\frac{e_i^{n+1}-e_i^n}{\Delta t}$ satisfies

$$\|\frac{\mathbf{e}^{n+1}-\mathbf{e}^n}{\Delta t}\|^2+(\nu_1+H_1)\Delta t\sum_{j=1}^{n}\|\nabla\left(\frac{e_1^{j+1}-e_1^j}{\Delta t}\right)\|^2$$
$$+(\nu_2+H_2)\Delta t\sum_{j=1}^{n}\|\nabla\left(\frac{e_2^{j+1}-e_2^j}{\Delta t}\right)\|^2\leq C\left(h^{2m}+(\Delta t)^2+H^2\right).$$

6.3.3 Correction Step Results

With this accuracy result under our belt, the stability and accuracy of the correction step are within reach. The next result seems to imply unconditional stability of the correction step; however, assumptions of Theorems 2 and 3 are needed to have this stability bound dependent only upon the problem data.

Theorem 4 (Stability of the Correction Step). *Let $\tilde{\mathbf{u}}^{n+1} \in \mathbf{X}^h$ satisfy (6.11) for each $n \in \{0, 1, 2, \cdots, \frac{T}{\Delta t} - 1\}$. Then $\exists C > 0$ independent of h, Δt such that $\tilde{\mathbf{u}}^{n+1}$ satisfies:*

$$
\begin{aligned}
&\left\|\tilde{u}_1{}^{n+1}\right\|^2 + \left\|\tilde{u}_2{}^{n+1}\right\|^2 + (\nu_1 + H_1)\Delta t \sum_{k=1}^{n+1} \left\|\nabla \tilde{u}_1{}^k\right\|^2 \\
&+ (\nu_2 + H_2)\Delta t \sum_{k=1}^{n+1} \left\|\nabla \tilde{u}_2{}^k\right\|^2 + \kappa\,\Delta t \int_I \left|\tilde{u}_1^{n+1}|[\tilde{\mathbf{u}}^n]|^{1/2} - \tilde{u}_2^n|[\tilde{\mathbf{u}}^{n-1}]|^{1/2}\right|^2 ds \\
&+ \kappa\,\Delta t \int_I \left|\tilde{u}_2^{n+1}|[\tilde{\mathbf{u}}^n]|^{1/2} - \tilde{u}_1^n|[\tilde{\mathbf{u}}^{n-1}]|^{1/2}\right|^2 ds \\
&\leq \frac{C\Delta t}{\nu_1 + H_1} \sum_{j=1}^{n} \Big[\|\nabla e_1^{j+1}\|^2 + \|e_2^j\|\|\nabla e_2^j\|\|\nabla e_i^j\|^2 \\
&+ \|\nabla e_i^{j+1}\|^2 + \|e_1^{j+1}\|\|\nabla e_1^{j+1}\|\|\nabla e_i^j\|^2 + \|e_1^j\|\|\nabla e_1^j\|\|\nabla e_i^j\|^2 + \|\nabla e_2^j\|^2 \Big] \\
&+ \frac{\Delta t}{14(\nu_1 + H_1)} \sum_{j=1}^{n} (\|\nabla e_2^j\|^2 + \|e_2^j\|\|\nabla e_2^j\|\|\nabla e_i^j\|^2) \\
&+ \frac{8\Delta t(\nu_1 + H_1)}{19} \sum_{j=1}^{n} \left\{ \Delta t^2 \|\nabla (\frac{e_1^{j+1} - e_1^j}{\Delta t})\|^2 + \Delta t^2 C^2_{\nabla \hat{u}_t}) \right\} \\
&+ \frac{8\Delta t(\nu_1 + H_1)}{19} \sum_{j=1}^{n} \Big[\Delta t \|\nabla \hat{u}_1^{j+1}\|^2 \Delta t \|\nabla (\frac{e_1^{j+1} - e_1^j}{\Delta t})\|^2 + \Delta t \|\nabla \hat{u}_1^{j+1}\|^2 \Delta t C^2_{\nabla \hat{u}_{1t}} \\
&+ \Delta t \|\nabla \hat{u}_1^j\|^2 \Delta t \|\nabla (\frac{e_1^{j+1} - e_1^j}{\Delta t})\|^2 + \Delta t \|\nabla \hat{u}_1^j\|^2 \Delta t C^2_{\nabla \hat{u}_{1t}} \Big] \\
&+ \frac{19\Delta t}{(\nu_1 + H_1)} \sum_{j=1}^{n} \Big[H_1^2 \|\nabla \hat{u}_1^{j+1}\|^2 + \|\frac{f_1^{j+1} + f_1^j}{2}\|^2_{-1} \Big] + \frac{\Delta t C_{\nabla u^{n+1}}}{\nu_1 + H_1} \sum_{j=1}^{n} \Big[1 + \kappa \|\nabla e_i^{j+1}\|^2 \Big] \\
&+ \Delta t C \sum_{j=1}^{n} (\|e_1^{j+1}\|^{1/2} \|\nabla e_1^{j+1}\|^{1/2} + \|e_2^{j+1}\|^{1/2} \|\nabla e_2^{j+1}\|^{1/2}) \|\nabla e_i^{j+1}\|^2 \qquad (6.16)
\end{aligned}
$$

Under the same restrictions the accuracy of the correction step solution is elevated from first to second order, in terms of the time step Δt and the artificial viscosity parameter H.

Theorem 5. *(Accuracy of Correction Step) Let the assumptions of Theorems 2 and 3 be satisfied. Then* $\exists C > 0$ *independent of* h, Δt *such that for any* $n \in \{0, 1, 2, \ldots, M-1 = \frac{T}{\Delta t} - 1\}$, *the solution* $\tilde{u}_i^{n+1}$ *of (6.11) satisfies*

$$\|\mathbf{u}^{n+1} - \tilde{\mathbf{u}}^{n+1}\|^2 + (\nu_1 + H_1)\Delta t \sum_{j=1}^{n+1} \|\nabla(u_1^j - \tilde{u}_1^j)\|^2 \tag{6.17}$$
$$+ (\nu_2 + H_2)\Delta t \sum_{j=1}^{n+1} \|\nabla(u_2^j - \tilde{u}_2^j)\|^2$$
$$\le C\left(h^4 + h^2\Delta t^2 + H_1^4 + H_1^2\Delta t^2 + H_2^4 + H_2^2\Delta t^2 + (\Delta t)^4\right).$$

TABLE 6.1 Accuracy of the defect step approximation $\hat{u}$

N	$\|\|u - \hat{u}\|\|_{L^2(0,T;L^2(\Omega))}$	rate	$\|\|u - \hat{u}\|\|_{L^2(0,T;H^1(\Omega))}$	rate
2	1.4798e-002	-	7.3869e-002	-
4	9.4941e-003	0.64	6.8654e-002	0.10
8	5.5097e-003	0.78	4.9680e-002	0.46
16	2.9407e-003	0.90	2.9957e-002	0.72
32	1.5262e-003	0.94	1.5786e-002	0.92
64	7.9193e-004	0.94	7.7512e-003	1.02
128	4.0867e-004	0.95	3.7515e-003	1.04

6.4 COMPUTATIONS

We will return to the fluid–fluid interaction problem in the future chapters, where a couple of qualitative tests will be introduced and discussed. Here we only present the quantitative test from [105], aimed at verifying the convergence rates.

The forcing and the boundary conditions are chosen so that the exact solution in the domain $\Omega = \Omega_1 \cup \Omega_2$ with $\Omega_1 = [0,1] \times [0,1]$ and $\Omega_2 = [0,1] \times [0,-1]$ is given by

$$u_{1,1} = a\nu_1 e^{-t} x^2(1-x)^2(1+y) + a e^{-t/2} x(1-x)\nu_1/\sqrt{\kappa a}$$
$$u_{1,2} = a\nu_1 e^{-t} xy(2+y)(1-x)(2x-1) + a e^{-t/2} y(2x-1)\nu_1/\sqrt{\kappa a}$$
$$u_{2,1} = a\nu_1 e^{-t} x^2(1-x)^2(1 + \frac{\nu_1}{\nu_2} y)$$
$$u_{2,2} = a\nu_1 e^{-t} xy(1-x)(2x-1)(2 + \frac{\nu_1}{\nu_2} y),$$

where $u_{i,j} : \Omega_i \to R, \forall i, j = 1, 2$. Take $a = 1$, $\nu_1 = 0.5$, $\nu_2 = 0.1$, $\kappa = 1$ and the final time $T = 1$.

Taylor-Hood finite elements were used in [105], the spatial mesh of diameter h was refined with $H_1 = H_2 = \Delta t = h$. The errors and convergence rates, computed in the tables below, verify the first order accuracy (with respect to H, Δt) of the defect step solution, and the second order accuracy of the correction step approximation.

TABLE 6.2 Accuracy of the correction step approximation $\tilde{u}$

N	$\|\|u - \tilde{u}\|\|_{L^2(0,T;L^2(\Omega))}$	rate	$\|\|u - \tilde{u}\|\|_{L^2(0,T;H^1(\Omega))}$	rate
2	1.0087e-002	-	6.0656e-002	-
4	5.1671e-003	0.96	4.2536e-002	0.51
8	2.3203e-003	1.15	2.3754e-002	0.84
16	8.7794e-004	1.40	1.0149e-002	1.22
32	2.8166e-004	1.64	3.3514e-003	1.60
64	8.1197e-005	1.79	9.2868e-004	1.85
128	2.2172e-005	1.87	2.3908e-004	1.96

The deferred correction procedure, introduced in this Chapter, can grow in various directions. In particular, it should be fairly straightforward to combine deferred correction with a multirate time stepping method, where different time scales would be used for the updates of the velocity in the air and in the sea. The dynamics of the atmospheric flows is usually faster than that of the ocean currents, so the multirate methods could be a good fit. We will return to this fluid–fluid formulation later, as we turn our attention to turbulence modeling.

CHAPTER 7

Large Eddy Simulation with Correction (LES-C)

A New Class of Turbulence Models

Previous chapters have prepared us for the idea that defect correction could be viewed as a numerical regularization of flows at high Reynolds numbers, or a means to reduce the temporal discretization error due in turbulent flows; it can even be viewed as a replacement of (part of) a turbulence model. In a sense, these chapters have gotten us warmed up to an idea that defect correction and turbulence modeling could go hand in hand. Starting from this chapter, the remainder of the book will be devoted to a method that proposes to treat existing turbulence models as defects, and use defect correction to improve the quality of the models' solution.

7.1 LES-C IDEA

The idea was first presented by the author in [143]. Even though the general approach could be implemented with any existing family of turbulence models, it has only been applied to LES models so far. The method in [143] sought to reduce the modeling error of an existing LES model via defect correction. The resulting family of models was called LES-C ("Lessi"), Large Eddy Simulation with Correction. One LES model to be improved by LES-C was chosen in [143]: the well known Approximate Deconvolution Model , [153], gave rise to the ADC – Approximate Deconvolution with Correction, via a two-step defect correction method. The method, applied to flows at high Reynolds numbers, seeks two consecutive approximation pairs (w_1, q_1), (w_2, q_2) of the Quantity of Interest (QoI). The QoI could be the velocity–pressure pair (u, p), the pair

DOI: 10.1201/9781003355977-7

of spatially filtered velocity $\bar{u}$ and pressure p, or any other pair of functions $(h(u), g(p))$.

Algorithm 7.1.1. *Turbulence Modeling with Correction.*

1. If needed, rewrite the NSE system so that the velocity–pressure pair (u, p) is transformed into the QoI pair $(h(u), g(p))$:

$$\widetilde{NSE}(QoI) = 0.$$

2. Using the turbulence model (TM), find the defect step approximation of the QoI:

$$TM(w_1, q_1) = 0.$$

3. The correction step provides a more accurate approximation of the QoI, via

$$TM(w_2, q_2) = TM(w_1, q_1) - \widetilde{NSE}(w_1, q_1).$$

Before we start the investigation (both theoretical and numerical) of the proposed LES-C models, let's try to predict the possible benefits of Algorithm 7.1.1. When properly executed, a given LES-C model should

- reduce the modeling error of the LES model;
- potentially reduce the temporal discretization error via deferred correction; when incorporated in the correction step, this joint defect-deferred correction approach could further reduce the overall error at no extra computational cost;
- resolve the possible bottlenecks in the decoupling methods – e.g., if the flows in the fluid–fluid interaction problem of the previous chapter are turbulent, then the chosen turbulence model would still suffer from the low order temporal accuracy of the partitioning method;
- be naturally parallelizable: the defect and correction steps of Algorithm 7.1.1 should be run on different cores in the "trapezoidal" manner, where each new correction step is one time level behind the previous correction (or defect) step; given that all these steps have the same structure of the left-hand side and only differ in the (known) right-hand side, the cores should complete the computations in almost the same amount of CPU time, and the idle time should be minimal.

The first step of the algorithm is not to be taken lightly. The success of the correction procedure hinges upon the understanding of the quantity of interest (QoI), that the chosen LES model aims to approximate. The LES model, discussed in [143], is the Approximate Deconvolution Model (ADM). The model seeks to approximate the solution pair $(\bar{u}, p)$, where p is the true pressure of the NSE, and $\bar{u}$ represents the spatially filtered velocity field.

7.2 IMPLEMENTING THE LES-C ALGORITHM: APPROXIMATE DECONVOLUTION WITH CORRECTION

The (spatial) filtering operator, used throughout this book and in [143], is the well known Germano differential filter, [79]. Given $\phi \in L^2_0(\Omega)$, $\overline{\phi} \in H^2(\Omega) \cap L^2_0(\Omega)$ is the unique solution of

$$A_\delta \overline{\phi} := -\delta^2 \Delta \overline{\phi} + \overline{\phi} = \phi \quad \text{in } \Omega, \tag{7.1}$$

subject to zero or periodic boundary conditions. Under such boundary conditions, this averaging operator commutes with differentiation. The filtering width δ marks the smallest resolved scale of the flow, typically of the size of the spatial mesh diameter.

Given enough regularity, the filtered velocity $\bar{u}$ converges quadratically to u as the filtering width δ approaches zero - see, e.g., Lemma 2 in Chapter 4, with $N = 0$. The well known accuracy results for the ADM (e.g., [82]) state that the model solution w is also within δ^2 from $\bar{u}$. Thus, from triangle inequality we get $\|w - u\| \le \|w - \bar{u}\| + \|\bar{u} - u\| \le O(\delta^2)$ and technically the ADM (at least the simplest member of the ADM family, discussed herein) could be also viewed as an approximation to the solution pair (u, p).

There are two key ingredients in making the LES-C models. One is to treat an LES model as "defect". Another idea is to recognize what this LES model aims to approximate. If we view the ADM as a model for approximating (u, p), then the NSE is not to be modified in any way in the first step of Algorithm 7.1.1. However, we will view the ADM as a model for the QoI $(\bar{u}, p)$. Thus, we start the derivation of our first LES-C model, Approximate Deconvolution with Correction (ADC), by replacing u with the equivalent expression $\bar{u} - \delta^2 \Delta \bar{u}$ in the NSE. The problem reads: find $(\bar{u}, \zeta, p) \in (X^h, X^h, Q^h)$ such that for any $(v^h, \xi^h, \chi^h) \in (X^h, X^h, Q^h)$

$$\begin{aligned} (\bar{u}_t, v^h) + \delta^2(\nabla \bar{u}_t, \nabla v^h) + \nu(\nabla \bar{u}, \nabla v^h) + \nu\delta^2(\nabla \zeta, \nabla v^h) \\ + b^*(\bar{u}, \bar{u}; v^h) - (p, \nabla \cdot v^h) = (f, v^h) \\ - \delta^2 b^*(\zeta, \bar{u}; v^h) - \delta^2 b^*(\bar{u}, \zeta; v^h) - \delta^4 b^*(\zeta, \zeta; v^h), \\ (\nabla \bar{u}, \nabla \xi^h) = (\zeta, \xi^h), \\ (\nabla \cdot \bar{u}, \chi^h) = 0. \end{aligned} \tag{7.2}$$

Comparing this formulation of the NSE with the ADM model (e.g., (4.3) in Chapter 4), we immediately notice, which terms in (7.2) have their counterparts in (4.3), and which ones do not.

One popular way to understand the actions of the ADM is to notice two extra terms in (4.3), compared to the NSE formulation for (u, p). These are regarded as the enhanced energy and enhanced energy dissipation rate (both increments being of size $O(\delta^2)$). However, comparing (4.3) to (7.2) we immediately see that their left-hand sides have the same structure, and the ADM "omits" the three terms in the right-hand side of (7.2). This is the "defect", to be corrected by the ADC.

With the first step of Algorithm 7.1.1 given by (7.2), we now formulate the ADM and ADC models. The ADM seeks $(w_1, \zeta_1, q_1) \in (X^h, X^h, Q^h)$ such that for any $(v^h, \xi^h, \chi^h) \in (X^h, X^h, Q^h)$

$$\begin{aligned}(w_{1,t}, v^h) + \delta^2(\nabla w_{1,t}, \nabla v^h) + \nu(\nabla w_1, \nabla v^h) + \nu\delta^2(\nabla \zeta_1, \nabla v^h) \\ +b^*(w_1, w_1; v^h) - (q_1, \nabla \cdot v^h) = (f, v^h), \\ (\nabla w_1, \nabla \xi^h) = (\zeta_1, \xi^h), \\ (\nabla \cdot w_1, \chi^h) = 0.\end{aligned} \tag{7.3}$$

The correction step of Algorithm 7.1.1 now reads: find $(w_2, \zeta_2, q_2) \in (X^h, X^h, Q^h)$ such that for any $(v^h, \xi^h, \chi^h) \in (X^h, X^h, Q^h)$

$$\begin{aligned}(w_{2,t}, v^h) + \delta^2(\nabla w_{2,t}, \nabla v^h) + \nu(\nabla w_2, \nabla v^h) + \nu\delta^2(\nabla \zeta_2, \nabla v^h) \\ +b^*(w_2, w_2; v^h) - (q_2, \nabla \cdot v^h) = (f, v^h) \\ -\delta^2 b^*(\zeta_1, w_1; v^h) - \delta^2 b^*(w_1, \zeta_1; v^h) - \delta^4 b^*(\zeta_1, \zeta_1; v^h), \\ (\nabla w_2, \nabla \xi^h) = (\zeta_2, \xi^h), \\ (\nabla \cdot w_2, \chi^h) = 0.\end{aligned} \tag{7.4}$$

This is the semi-discrete (continuous in time) formulation of the ADC, convenient for understanding how the LES-C is constructed. For the numerical analysis we will need to discretize (7.3)–(7.4) in time. Following [143], we use the Backward Euler method; it results in the equations for $(w_1^{n+1}, \zeta_1^{n+1}, q_1^{n+1})$, $(w_2^{n+1}, \zeta_2^{n+1}, q_2^{n+1}) \in (X^h, X^h, Q^h), \forall (v^h, \xi^h, \chi^h) \in (X^h, X^h, Q^h)$ at $t = t_{n+1}$, $n \geq 0$, with $\Delta t := t_{i+1} - t_i$:

$$\begin{aligned}(\frac{w_1^{n+1} - w_1^n}{\Delta t}, v^h) + \delta^2(\frac{\nabla w_1^{n+1} - \nabla w_1^n}{\Delta t}, \nabla v^h) + \nu(\nabla w_1^{n+1}, \nabla v^h) \\ +\nu\delta^2(\nabla \zeta_1^{n+1}, \nabla v^h) + b^*(w_1^{n+1}, w_1^{n+1}; v^h) - (q_1^{n+1}, \nabla \cdot v^h) \\ = (f(t_{n+1}), v^h), (\nabla w_1^{n+1}, \nabla \xi^h) = (\zeta_1^{n+1}, \xi^h), \\ (\nabla \cdot w_1^{n+1}, \chi^h) = 0.\end{aligned} \tag{7.5}$$

$$\begin{aligned}(\frac{w_2^{n+1} - w_2^n}{\Delta t}, v^h) + \delta^2(\frac{\nabla w_2^{n+1} - \nabla w_2^n}{\Delta t}, \nabla v^h) + \nu(\nabla w_2^{n+1}, \nabla v^h) \\ +\nu\delta^2(\nabla \zeta_2^{n+1}, \nabla v^h) + b^*(w_2^{n+1}, w_2^{n+1}; v^h) - (q_2^{n+1}, \nabla \cdot v^h) \\ = (f(t_{n+1}), v^h) - \delta^2 b^*(\zeta_1^{n+1}, w_1^{n+1}; v^h) - \delta^2 b^*(w_1^{n+1}, \zeta_1^{n+1}; v^h) \\ -\delta^4 b^*(\zeta_1^{n+1}, \zeta_1^{n+1}; v^h), (\nabla w_2^{n+1}, \nabla \xi^h) = (\zeta_2^{n+1}, \xi^h), \\ (\nabla \cdot w_2^{n+1}, \chi^h) = 0.\end{aligned} \tag{7.6}$$

7.3 NUMERICAL ANALYSIS OF THE ADC

The following norm, induced by the deconvolution operator A, is used in [143]:

$$\|\phi\|_A^2 = \|\phi\|^2 + \delta^2\|\nabla\phi\|^2.$$

7.3.1 ADM-Related Results

Stability and accuracy of the ADM model (7.5) are well-established – see, e.g., the following theorems from [57].

Theorem 14 (Stability of the first approximation). *Let w_1 satisfy (7.5). Let $f \in L^2(0,T;H^{-1}(\Omega))$. Then, for $n = 0,\dots,N-1$,*

$$\|w_1^{n+1}\|_A^2 + \nu\Delta t\Sigma_{i=0}^n\|\nabla w_1^{i+1}\|^2 + \nu\delta^2\Delta t\Sigma_{i=0}^n\|\zeta_1^{i+1}\|^2 \\ \le \|w_1^0\|_A^2 + \frac{1}{\nu}\Delta t\Sigma_{i=0}^n\|f(t_{i+1})\|_{-1}^2.$$

Theorem 15 (Accuracy of the first approximation). *Let the time step satisfy*

$$\Delta t < \frac{\nu^3}{\max_{i=0,1,..,N}\|\nabla\bar{u}(t_i)\|^4}. \tag{7.7}$$

Let also $\bar{u} \in L^2(0,T;H^3(\Omega))$ and $\bar{u}_{tt} \in L^2(0,T;H^1(\Omega))$. The error in the first approximation satisfies

$$\|\bar{u}(t_{n+1}) - w_1^{n+1}\|_A^2 + \nu\Delta t\sum_{i=0}^n\|\nabla(\bar{u}(t_{i+1}) - w_1^{i+1})\|^2 \tag{7.8}$$

$$+\nu\delta^2\Delta t\sum_{i=0}^n\|\zeta(t_{i+1}) - \zeta_1^{i+1}\|^2 \le C(\Delta t^2 + \delta^4$$

$$+\Delta t\sum_{i=0}^n((1+\delta^2h^{-2})\inf_{v\in V^h}\|\nabla(\bar{u}(t_i) - v^i)\|^2 + \inf_{q\in Q^h}\|p(t_i) - q^i\|^2)).$$

7.3.2 Accuracy of the Gradient of the Defect Step Error

The reader might not be surprised by the need to establish an accuracy result for the (spatial) derivatives of the error; after all, the extra terms in the right-hand side of (7.6) involve higher derivatives of the ADM velocity w_1. It requires extra accuracy assumptions on the true solution $\bar{u}$.

Theorem 16 (Accuracy of the Laplacian). *Let the assumptions of Theorem 15 be satisfied. Let also $\bar{u} \in L^\infty(0,T;L^\infty(\Omega))$, $\nabla\bar{u} \in L^\infty(0,T;L^\infty(\Omega))$, $\Delta\bar{u} \in L^2(0,T;H^1(\Omega))\cap L^\infty(0,T;L^\infty(\Omega))$, $\nabla\bar{u}_{tt} \in L^2(0,T;H^1(\Omega))$. Let the time step and the filtering width satisfy $\Delta t \le h^{1/2}, \delta \le h^{1/4}$. Then the following accuracy result holds for the solution $(w_1^{n+1}, \zeta_1^{n+1}, q_1^{n+1})$ of (7.5):*

$$\|\nabla(\bar{u}(t_{n+1}) - w_1^{n+1})\|^2 + \delta^2\|\zeta(t_{n+1}) - \zeta_1^{n+1}\|^2 \tag{7.9}$$

$$+\nu\Delta t\sum_{i=0}^{n}\|\varsigma(t_{i+1})-\varsigma_1^{i+1}\|^2+\nu\delta^2\Delta t\sum_{i=0}^{n}\|\nabla(\varsigma(t_{i+1})-\varsigma_1^{i+1})\|^2$$
$$\leq C(\Delta t^2+\Delta t^2\delta^2+\delta^4+\Delta t\sum_{i=0}^{n}(\inf_{v\in V^h}\|\nabla(\bar{u}(t_i)-v^i)\|^2$$
$$+\inf_{r\in V^h}\|\varsigma(t_i)-r^i\|^2+\min(\delta^{-2},h^{-2})\cdot\inf_{q\in Q^h}\|p(t_i)-q^i\|^2)).$$

7.3.3 Stability and Accuracy of the ADC

The stability and accuracy results for the correction step are now within reach.

Theorem 17 (Stability of the ADC). *Let w_2 satisfy (7.6). Let the assumptions of Theorem 16 be satisfied; let also $\Delta\bar{u}\in L^\infty(0,T;H^1(\Omega))$. Then there exists C independent of $h,\Delta t$ such that, for $n=0,\ldots,N-1$,*

$$\|w_2^{n+1}\|_A^2+\nu\Delta t\Sigma_{i=0}^n\|\nabla w_2^{i+1}\|^2+\nu\delta^2\Delta t\Sigma_{i=0}^n\|\varsigma_2^{i+1}\|^2$$
$$\leq\frac{C}{\nu^2}\left(\|w_2^0\|_A^2+\frac{1}{\nu}\Delta t\Sigma_{i=0}^n\|f(t_{i+1})\|_{-1}^2\right).$$

Theorem 18 (Accuracy of the ADC). *Let the assumptions of Theorem 16 be satisfied. Let also $\nabla(\Delta\bar{u})\in L^\infty(0,T;L^\infty(\Omega))$. Then the accuracy of the correction step approximation is given by*

$$\|\bar{u}(t_{n+1})-w_2^{n+1}\|_A^2+\nu\Delta t\sum_{i=0}^{n}\|\nabla(\bar{u}(t_{i+1})-w_2^{i+1})\|^2 \tag{7.10}$$
$$+\nu\delta^2\Delta t\sum_{i=0}^{n}\|\varsigma(t_{i+1})-\varsigma_2^{i+1}\|^2\leq C(\Delta t^2+\Delta t^2\delta^4+\delta^8$$
$$+\Delta t\sum_{i=0}^{n}((1+\nu\delta^2h^{-2})\inf_{v^h\in V^h}\|\nabla(\bar{u}(t_i)-v^h)\|^2$$
$$+\inf_{r^h\in V^h}\|\varsigma(t_i)-r^h\|^2+\inf_{q^h\in Q^h}\|p(t_i)-q^h\|^2)).$$

Compared to the ADM, the modeling error of the ADC is improved from $O(\delta^2)$ to $O(\delta^4)$. The fact that this modeling error is only one of the three components of the overall error, is one of the main challenges in validating the LES-C models. Indeed, equations (7.5)–(7.6) do not aim to improve the accuracy due to spatial or temporal discretization. The latter will be addressed in the next chapter. We now proceed to the discussion on numerical validation of the ADC.

7.4 NUMERICAL VALIDATION OF THE ADC

The numerical validation of the ADC model (7.5)–(7.6) was presented in [143]. In addition to checking the convergence rates for a problem with a manufactured solution, several qualitative tests were provided, including the computation of drag and lift coefficients for a flow past a circular obstacle. One interesting analytic result that helps to understand the action of the ADC upon the small scales of the velocity field is presented via the Taylor–Green vortex problem. Presented e.g., in [154, 157, 155, 156] this problem discusses the flow in a 2π-periodic box with zero forcing. We are mainly interested in this analytic test, because of the closed form of the Taylor solutions of the NSE: it was shown in [156] that the ADM solutions of this problem have similar structure, so it is of interest to see what the ADC could bring to the table.

7.4.1 Taylor–Green Vortex

Define the Green–Taylor solutions via

Definition 2. *Let $\phi(x)$ be 2π-periodic and satisfy*

$$-\Delta\phi = \lambda\phi, \ \nabla\cdot\phi = 0.$$

Then the velocity–pressure pair (u, p) is a solution of the NSE, where

$$\begin{aligned} u(x,t) &= e^{-\nu\lambda t}\phi(x), \\ u\cdot\nabla u + \nabla p &= 0. \end{aligned} \tag{7.11}$$

The results for the ADM and the ADC solution pairs (w_1, p_1), (w_2, p_2) are given in the theorem below. The ADM part was proven in [156], and the ADC result comes from [143].

Theorem 19. *There exist solution pairs (w_1, p_1) of the ADM and (w_2, p_2) of the ADC, respectively, where*

$$\begin{aligned} w_1(x,t) &= e^{-\nu\lambda t}\phi(x), \\ w_1\cdot\nabla w_1 + \nabla p_1 &= 0, \end{aligned} \tag{7.12}$$

and

$$\begin{aligned} w_2(x,t) &= e^{-\nu\lambda t}\phi(x), \\ w_2\cdot\nabla w_2 + \nabla(p_2 - ((1+\lambda\delta^2)^2 - 1)p_1) &= 0. \end{aligned} \tag{7.13}$$

When comparing the structure of these strong solutions of the fully continuous formulations of the ADM/ADC, [143], the role of the extra pressure terms in the ADC is unclear. It helps to remember that the models aim to approximate the filtered velocity $\bar{u}$ rather than u. The filtered NSE solution for the Taylor problem is presented in the theorem below, [143].

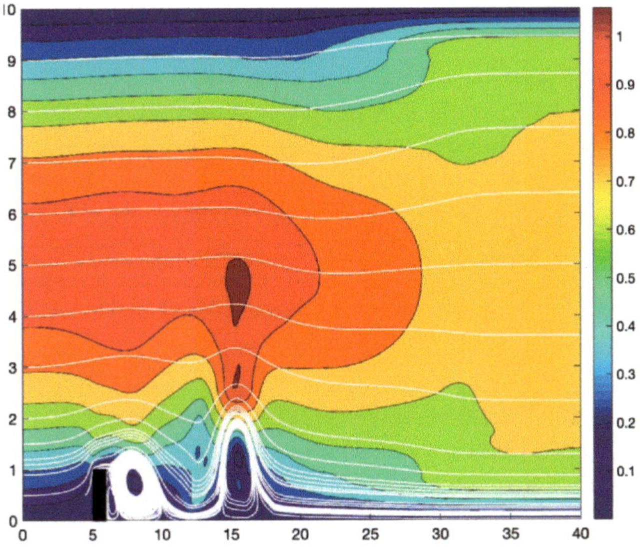

Figure 7.1 Filtered true velocity, $Ndofs = 33,820$.

Theorem 20. *A solution pair* $(\bar{u}, p)$ *of the filtered NSE exists, where*

$$\bar{u}(x,t) = e^{-\nu\lambda t}\phi(x), \tag{7.14}$$
$$(1+\lambda\delta^2)^2\bar{u}\cdot\nabla\bar{u} + \nabla p = 0.$$

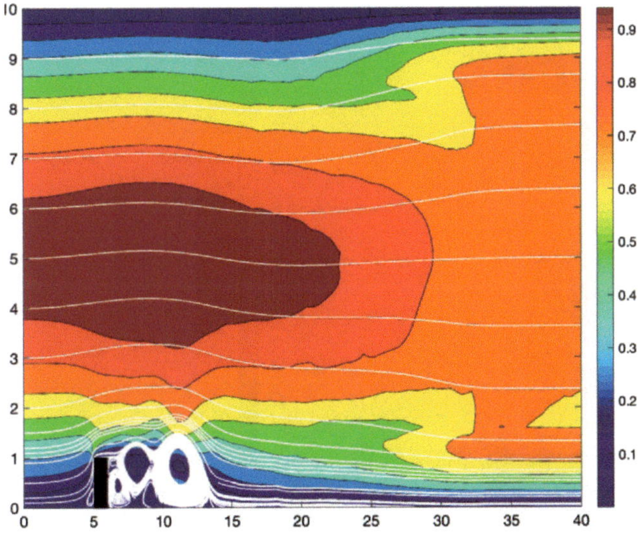

Figure 7.2 ADM solution, $Ndofs = 8897$.

Now, comparing all three of the above solutions, we get a clearer picture.

$$\begin{aligned} \bar{u}\cdot\nabla\bar{u} + (2\lambda\delta^2 + \lambda^2\delta^4)\bar{u}\cdot\nabla\bar{u} + \nabla p = 0, \\ w_1\cdot\nabla w_1 + \nabla p_1 = 0, \\ w_2\cdot\nabla w_2 + (2\lambda\delta^2 + \lambda^2\delta^4)w_1\cdot\nabla w_1 + \nabla p_2 = 0. \end{aligned} \tag{7.15}$$

At large scales (small values of λ) the models seem comparable. It is the small scales where the ADC aims to outperform the ADM. The "extra" pressure terms in (7.13) have the structure of the small-scale terms that are present in the (filtered) NSE true solution, but are completely missed by the ADM.

7.4.2 Flow Past a Step

A 2-D flow past a forward-backward facing step is the benchmark problem that we will investigate next. This is a well-known problem, studied, e.g., in [88, 160, 161, 78]. At a certain range of the Reynolds numbers, $500 \le Re \le 700$, the flow contains a set of coherent structures (vortices) that form, travel and shed. This is happening at moderate Reynolds numbers, so one could compute a solution without a turbulence model, using a chosen finite element discretization on a fine enough mesh with a small enough time step – and consider this solution to be the reference solution.

As can be seen from the next plot, the ADM failed to predict the correct shedding, detachment and traveling of the vortices. This was computed at a much coarser spatial mesh; refining the mesh and reducing the filtering width δ should improve the quality of the ADM solution – but the price is

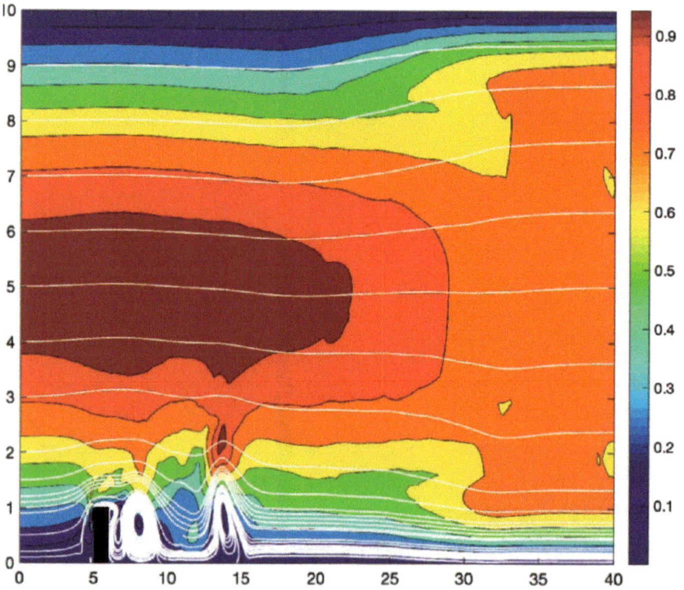

Figure 7.3 ADC solution, $Ndofs = 8897$.

the increased size and complexity of the system and therefore, a much longer CPU time.

On the other hand, utilizing the same coarse spatial mesh the ADC produces a much higher quality solution, that correctly captures the shedding and detachment, even though the vortex is not traveling as fast as it should.

The numerical tests of this Chapter (and some more tests from [143]) clearly indicate the overall improvement of the ADC model from the LES-C family over its LES counterpart, the ADM. The term used by the author is "accuracy per computational cost": the ADC can produce a higher accuracy solution given the same computational resources as the ADM (plus one extra core for the parallel computation of the correction step); or the ADC can get within the prescribed tolerance from the reference solution faster than the ADM. Of course, these results should be regarded as preliminary, to be followed by testing these models in a turbulent setting.

7.5 SOURCE CODE

The following FreeFEM++ code was used in modeling the flow past a step.

```
// Flow past a step, modeled by the Approximate Deconvolution
    with Correction.
// Crank-Nicolson method is used for time discretization.
// Resulting velocity field could be passed on to a plotting
    software.

real s0=clock();

border A(t=0,5.0){x=t; y=0.0; label=1;};
border B(t=0,1.0){x=5.0; y=t; label=2;};
border C(t=0,1.0){x=5.0+t; y=1.0; label=3;};
border D(t=0,1.0){x=6.0; y=1.0-t; label=4;};
border E(t=0,7.0){x=6.0+t; y=0.0; label=5;};
border F(t=0,87.0){x=13.0+t; y=0.0; label=6;};
border G(t=0,10.0){x=100.0; y=t; label=7;};
border H(t=0,100.0){x=100.0-t; y=10.0; label=8;};
border K(t=0,10.0){x=0.0; y=10.0-t; label=9;};



int N=20;
real h=1.0/N;

real delta = 1.0/8.0; //filtering width

real gamma = 1.0;// grad-div stabilization parameter, that is
    needed due to the weak enforcement
//of the divergence-free condition by the Taylor-Hood finite
    element pair.
```

```
mesh Th=buildmesh( A(N)+B(N)+C(N)+D(N)+E(N)+F(3*N)+G(N/2)+H(3*N)+
    K(N) );

fespace Vh2(Th,P2);
fespace Vh(Th,P1);

Vh2 u2,v2,up1,up2,u2old,um2,e2,e22, bd1, bd2, cum1, cum2, u1half,
     u2half, bdDzeta1, bdDzeta2;
Vh bdp, q2, qlambda1, qlambda2, lambda1, lambda2;
Vh2 u1, v1, u1old, um1, e1, v3, v4, cu1old, cu2old;
Vh2 dcvel1, dcvel2, dcvel1old, dcvel2old, dcvelm1, dcvelm2, dce1,
     dce2;
Vh p, pm, q, e3, pold, dcp, dcpold, dce3, dcpm, cp, cpm, cpold;
Vh2 true1, true2;


//Vh2 TrueFilteredVelTh1, TrueFilteredVelTh2, TrueDzeta1,
    TrueDzeta2;

//Vh2 dxmixDzetaPrev1, dxmixDzetaPrev2, dymixDzetaPrev1,
    dymixDzetaPrev2, dxmixDzeta1, dxmixDzeta2, dymixDzeta1,
    dymixDzeta2;

// *********mixed formulation variables****
Vh2 mixDzeta1,mixDzeta2,mixv1,mixv2 ,mixDzeta1half, mixDzeta2half
    ;
Vh2 mixDCDzeta1,mixDCDzeta2,mixv3,mixv4;
Vh2 mixDzetaPrev1,mixDzetaPrev2, mixcDzetaPrev1,mixcDzetaPrev2;
Vh2 MixedBoundaryCond1,MixedBoundaryCond2, cu1,cu2,mixcDzeta1,
    mixcDzeta2;
// ********************************************

Vh psi, phi, speedU, speedCU;

real reynolds=600.0;
real nu = 1.0/reynolds;
real err1, err2, dcerr1;

real T=40;
real dt=0.01;
int nbiter = T/dt;
int itnl=0,iter=0, MaxNlIt=30;
real t;


func real ParabolicInflow (real t) {
return y*(10-y)/25;
}

func real NegLaplacianParabolicInflow (real t) {
return 2.0/25;
}


func real truevel1 (real t) {
return 0.0;
}
```

```
func real truevel2 (real t) {
return 0.0;
}

func real truepres (real t) {
return (40-x)/40;
}

func real f1 (real t) {
return 0.0;
}

func real f2 (real t) {
return 0.0;
}


problem ADM ([u1,u2,mixDzeta1,mixDzeta2,p,lambda1],[v1,v2,mixv1,
    mixv2,q,qlambda1],solver=UMFPACK) =
    int2d(Th)(
              (1.0/dt)*( u1*v1 + u2*v2)

            + (1.0/dt)*(delta^2)*( dx(u1)*dx(v1) + dy(u1)*dy(v1)
                                + dx(u2)*dx(v2) + dy(u2)*dy(v2) )


            +0.125*( u1old*dx(u1)*v1 + u2old*dy(u1)*v1
              + u1old*dx(u2)*v2 + u2old*dy(u2)*v2 )
            -0.125*( u1old*dx(v1)*u1 + u2old*dy(v1)*u1    //b*(
                u_old,u,v)
              + u1old*dx(v2)*u2 + u2old*dy(v2)*u2 )


            +0.125*( u1*dx(u1old)*v1 + u2*dy(u1old)*v1
              + u1*dx(u2old)*v2 + u2*dy(u2old)*v2 )
            -0.125*( u1*dx(v1)*u1old + u2*dy(v1)*u1old    //b*(u,
                u_old,v)
              + u1*dx(v2)*u2old + u2*dy(v2)*u2old )


            +0.125*( um1*dx(u1)*v1 + um2*dy(u1)*v1
              + um1*dx(u2)*v2 + um2*dy(u2)*v2 )
            -0.125*( um1*dx(v1)*u1 + um2*dy(v1)*u1    //b*(u_prev,
                u,v)
              + um1*dx(v2)*u2 + um2*dy(v2)*u2 )


            +0.125*( u1*dx(um1)*v1 + u2*dy(um1)*v1
              + u1*dx(um2)*v2 + u2*dy(um2)*v2 )
            -0.125*( u1*dx(v1)*um1 + u2*dy(v1)*um1    //b*(u,
                u_prev,v)
              + u1*dx(v2)*um2 + u2*dy(v2)*um2 )
```

```
          + 0.5*nu * ( dx(u1)*dx(v1) + dy(u1)*dy(v1)
          +            dx(u2)*dx(v2) + dy(u2)*dy(v2) )

          + 0.5*nu*(delta^2)* ( dx(mixDzeta1)*dx(v1) + dy(
              mixDzeta1)*dy(v1)
          +            dx(mixDzeta2)*dx(v2) + dy(mixDzeta2)*dy(v2)
              )

          + p*q*(0.000000001)
          - 0.5*p*(dx(v1) + dy(v2))
          + (dx(u1) + dy(u2))*q
          + gamma*(dx(u1)+dy(u2))*(dx(v1)+dy(v2)) //grad-div
              stabilization term
        )
+ int2d(Th) (
          - (1.0/dt)*( um1*v1 + um2*v2)

          - (1.0/dt)*(delta^2)*( dx(um1)*dx(v1) + dy(um1)*dy(v1
              )
          +            dx(um2)*dx(v2) + dy(um2)*dy(v2) )

           -0.125*( u1old*dx(u1old)*v1 + u2old*dy(u1old)*v1
            + u1old*dx(u2old)*v2 + u2old*dy(u2old)*v2 )
          +0.125*( u1old*dx(v1)*u1old + u2old*dy(v1)*u1old     //
               -b*(u_old,u_old,v)
            + u1old*dx(v2)*u2old + u2old*dy(v2)*u2old )

          +0.125*( um1*dx(um1)*v1 + um2*dy(um1)*v1
            + um1*dx(um2)*v2 + um2*dy(um2)*v2 )
          -0.125*( um1*dx(v1)*um1 + um2*dy(v1)*um1     //b*(
              u_prev,u_prev,v)
            + um1*dx(v2)*um2 + um2*dy(v2)*um2 )


          + 0.5*nu * ( dx(um1)*dx(v1) + dy(um1)*dy(v1)
          +            dx(um2)*dx(v2) + dy(um2)*dy(v2) )

          + 0.5*nu*(delta^2)* ( dx(mixDzetaPrev1)*dx(v1) + dy(
              mixDzetaPrev1)*dy(v1)
          +            dx(mixDzetaPrev2)*dx(v2) + dy(mixDzetaPrev2)
              *dy(v2) )

          - 0.5*pm*(dx(v1) + dy(v2))

          - 0.5*(f1(t)*v1 + f2(t)*v2)
          - 0.5*(f1(t-dt)*v1 + f2(t-dt)*v2)
           )
+ int2d(Th) (
           ( mixDzeta1*mixv1 + mixDzeta2*mixv2)
           - ( dx(u1)*dx(mixv1) + dy(u1)*dy(mixv1)
             +            dx(u2)*dx(mixv2) + dy(u2)*dy(mixv2) )
           - lambda1*(dx(mixv1)+dy(mixv2))
           + (dx(mixDzeta1)+dy(mixDzeta2))*qlambda1
```

```
            + gamma*(dx(mixDzeta1)+dy(mixDzeta2))*(dx(mixv1)+dy(
                mixv2)) //grad-div stabilization term
            )
    + on(1,2,3,4,5,6,8, u1=0,u2=0,mixDzeta1=0,mixDzeta2=0)
    + on(7, u1=bd1,u2=0,mixDzeta1=NegLaplacianParabolicInflow(t),
        mixDzeta2=0,p=0)
    + on(9, u1=bd1,u2=0,mixDzeta1=NegLaplacianParabolicInflow(t),
        mixDzeta2=0);


problem ADC ([cu1,cu2,mixcDzeta1,mixcDzeta2,cp, lambda2],[v3,v4,
    mixv3,mixv4,q2, qlambda2],solver=UMFPACK) =
    int2d(Th)(
            (1.0/dt)*( cu1*v3 + cu2*v4)

            + (1.0/dt)*(delta^2)*( dx(cu1)*dx(v3) + dy(cu1)*dy(v3
                )
                                + dx(cu2)*dx(v4) + dy(cu2)*dy(v4)
                                    )


            +0.125*( cu1old*dx(cu1)*v3 + cu2old*dy(cu1)*v3
              + cu1old*dx(cu2)*v4 + cu2old*dy(cu2)*v4 )
            -0.125*( cu1old*dx(v3)*cu1 + cu2old*dy(v3)*cu1     //b
                *(cu_old,cu,v)
              + cu1old*dx(v4)*cu2 + cu2old*dy(v4)*cu2 )


            +0.125*( cu1*dx(cu1old)*v3 + cu2*dy(cu1old)*v3
              + cu1*dx(cu2old)*v4 + cu2*dy(cu2old)*v4 )
            -0.125*( cu1*dx(v3)*cu1old + cu2*dy(v3)*cu1old     //b
                *(cu,cu_old,v)
              + cu1*dx(v4)*cu2old + cu2*dy(v4)*cu2old )


            +0.125*( cum1*dx(cu1)*v3 + cum2*dy(cu1)*v3
              + cum1*dx(cu2)*v4 + cum2*dy(cu2)*v4 )
            -0.125*( cum1*dx(v3)*cu1 + cum2*dy(v3)*cu1     //b*(
                cu_prev,cu,v)
              + cum1*dx(v4)*cu2 + cum2*dy(v4)*cu2 )


            +0.125*( cu1*dx(cum1)*v3 + cu2*dy(cum1)*v3
              + cu1*dx(cum2)*v4 + cu2*dy(cum2)*v4 )
            -0.125*( cu1*dx(v3)*cum1 + cu2*dy(v3)*cum1     //b*(cu,
                cu_prev,v)
              + cu1*dx(v4)*cum2 + cu2*dy(v4)*cum2 )




            + 0.5*nu * ( dx(cu1)*dx(v3) + dy(cu1)*dy(v3)
            +           dx(cu2)*dx(v4) + dy(cu2)*dy(v4) )

            + 0.5*nu*(delta^2)* ( dx(mixcDzeta1)*dx(v3) + dy(
                mixcDzeta1)*dy(v3)
```

```
            +        dx(mixcDzeta2)*dx(v4) + dy(mixcDzeta2)*dy(v4
                  ) )


            + cp*q2*(0.000000001)

            - 0.5*cp*(dx(v3) + dy(v4))

            + (dx(cu1) + dy(cu2))*q2

            + gamma*(dx(cu1)+dy(cu2))*(dx(v3)+dy(v4)) //grad-div
                  stabilization term
          )
  + int2d(Th) (
            - (1.0/dt)*( cum1*v3 + cum2*v4)

            - (1.0/dt)*(delta^2)*( dx(cum1)*dx(v3) + dy(cum1)*dy(
                  v3)
            +        dx(cum2)*dx(v4) + dy(cum2)*dy(v4) )

            -0.125*( cu1old*dx(cu1old)*v3 + cu2old*dy(cu1old)*v3
              + cu1old*dx(cu2old)*v4 + cu2old*dy(cu2old)*v4 )
            +0.125*( cu1old*dx(v3)*cu1old + cu2old*dy(v3)*cu1old
                  // -b*(cu_old, cu_old, v)
              + cu1old*dx(v4)*cu2old + cu2old*dy(v4)*cu2old )

            +0.125*( cum1*dx(cum1)*v3 + cum2*dy(cum1)*v3
              + cum1*dx(cum2)*v4 + cum2*dy(cum2)*v4 )
            -0.125*( cum1*dx(v3)*cum1 + cum2*dy(v3)*cum1    //b*(
                  cu_prev, cu_prev, v)
              + cum1*dx(v4)*cum2 + cum2*dy(v4)*cum2 )


            + 0.5*nu * ( dx(cum1)*dx(v3) + dy(cum1)*dy(v3)
            +        dx(cum2)*dx(v4) + dy(cum2)*dy(v4) )

            + 0.5*nu*(delta^2)* ( dx(mixcDzetaPrev1)*dx(v3) + dy(
                  mixcDzetaPrev1)*dy(v3)
            +        dx(mixcDzetaPrev2)*dx(v4) + dy(
                  mixcDzetaPrev2)*dy(v4) )

            - 0.5*cpm*(dx(v3) + dy(v4))

            - 0.5*(f1(t)*v3 + f2(t)*v4)
            - 0.5*(f1(t-dt)*v3 + f2(t-dt)*v4)

//**** Now the three extra terms for spatial correction of the
      turbulence model **

            +0.125*(delta^2)*( mixDzeta1half*dx(u1half)*v3 +
                  mixDzeta2half*dy(u1half)*v3
              + mixDzeta1half*dx(u2half)*v4 + mixDzeta2half*dy
                  (u2half)*v4 )

            -0.125*(delta^2)*( mixDzeta1half*dx(v3)*u1half +
                  mixDzeta2half*dy(v3)*u1half
```

```
                + mixDzeta1half*dx(v4)*u2half + mixDzeta2half*dy
                    (v4)*u2half )



            +0.125*(delta^2)*( u1half*dx(mixDzeta1half)*v3 +
                u2half*dy(mixDzeta1half)*v3
                + u1half*dx(mixDzeta2half)*v4 + u2half*dy(
                    mixDzeta2half)*v4 )

            -0.125*(delta^2)*( u1half*dx(v3)*mixDzeta1half +
                u2half*dy(v3)*mixDzeta1half
                + u1half*dx(v4)*mixDzeta2half + u2half*dy(v4)*
                    mixDzeta2half )


            +0.125*(delta^4)*( mixDzeta1half*dx(mixDzeta1half)*v3
                +  mixDzeta2half*dy(mixDzeta1half)*v3
                + mixDzeta1half*dx(mixDzeta2half)*v4 +
                    mixDzeta2half*dy(mixDzeta2half)*v4 )

            -0.125*(delta^4)*( mixDzeta1half*dx(v3)*mixDzeta1half
                + mixDzeta2half*dy(v3)*mixDzeta1half
                + mixDzeta1half*dx(v4)*mixDzeta2half +
                    mixDzeta2half*dy(v4)*mixDzeta2half )


//*************************************************

             )
  + int2d(Th) (
             ( mixcDzeta1*mixv3 + mixcDzeta2*mixv4)
             - ( dx(cu1)*dx(mixv3) + dy(cu1)*dy(mixv3)
               +         dx(cu2)*dx(mixv4) + dy(cu2)*dy(mixv4) )
             - lambda2*(dx(mixv3)+dy(mixv4))
             + (dx(mixcDzeta1)+dy(mixcDzeta2))*qlambda2

            + gamma*(dx(mixcDzeta1)+dy(mixcDzeta2))*(dx(mixv3)+dy
                (mixv4)) //grad-div stabilization term
             )
  + on(1,2,3,4,5,6,8, cu1=0,cu2=0,mixcDzeta1=0,mixcDzeta2=0)
  + on(7, cu1=bd1,cu2=0,mixcDzeta1=NegLaplacianParabolicInflow(t)
      ,mixcDzeta2=0,cp=0)
  + on(9, cu1=bd1,cu2=0,mixcDzeta1=NegLaplacianParabolicInflow(t)
      ,mixcDzeta2=0);

t=0.0;

u1 = 0;
u2 = 0;
p = 0;

lambda1=0;

cu1 = 0;
cu2 = 0;
cp = 0;
```

```
lambda2=0;



mixDzeta1 = 0;
mixDzeta2 = 0;




mixcDzeta1 = 0;
mixcDzeta2 = 0;



for (iter=1;iter<=nbiter; iter++)
{
   um1 = u1;
   um2 = u2;
   u1old = u1;
   u2old = u2;
   pold = p;
   pm=p;

   mixDzetaPrev1 = mixDzeta1;
   mixDzetaPrev2 = mixDzeta2;


   cum1 = cu1;
   cum2 = cu2;
   cu1old = cu1;
   cu2old = cu2;
   cpold = cp;
   cpm=cp;

   mixcDzetaPrev1 = mixcDzeta1;
   mixcDzetaPrev2 = mixcDzeta2;


   t = t + dt;

   bd1=ParabolicInflow(t);


   err1 = 1;
   dcerr1 = 1;
   cout << " iter = " << iter << " ------------------------------------ " <<
       endl;

   for (itnl=1; itnl<=MaxNlIt; itnl++)
   {
       cout<< "itnl = " << itnl << endl ;

       ADM;
       e1 = abs(u1 - u1old);
       e2 = abs(u2 - u2old);
```

```
        cout<< "e1[].max " << e1[].max << "      e2[].max " << e2
            [].max<< endl ;
        cout<< " " << endl ;
        if ( (e1[].max < 0.0001) && (e2[].max < 0.0001) )
         {
            break ;
         }
         u1old = u1 ;
         u2old = u2 ;
//         pold = p;
         err1 = err1 + 1 ;
    }

    if ( err1 >= MaxNlIt )
    {
       cout << "approx failed for iter = " << iter << endl ;
       break;
    }

    //———————————————————————————————


// ------Now the Correction step----


    u1half = u1+um1;
    u2half = u2+um2;

    mixDzeta1half = mixDzeta1+mixDzetaPrev1;
    mixDzeta2half = mixDzeta2+mixDzetaPrev2;


    cout << " iter = " << iter << " ——————————————————— " <<
        endl;

    for (itnl=1; itnl<=MaxNlIt; itnl++)
    {
        cout<< "itnl = " << itnl << endl ;

        ADC;
        dce1 = abs(cu1 - cu1old);
        dce2 = abs(cu2 - cu2old);

        cout << "       dce1[].max "<< dce1[].max << "       dce2[].
            max " << dce2[].max << endl ;
        cout<< " " << endl ;
        if ( (dce1[].max < 0.0001) && (dce2[].max < 0.0001) )
         {
            break ;
         }
         cu1old = cu1 ;
         cu2old = cu2 ;
         cpold = cp;
         dcerr1 = dcerr1 + 1 ;
    }
```

```
    if ( dcerr1 >= MaxNlIt )
    {
        cout << "approx failed for iter = " << iter << endl ;
        break;
    }
}

cout << "CPU " << clock()-s0 << "s " << endl;
```

CHAPTER 8

Scale Truncation and the Energy Balance of the ADC

The investigation of the ADC model (7.3)–(7.4) continues with the discussion on scale truncation and energy balance. Adding to our understanding of the models' actions on the small scales, we will consider the known results of [77] for the ADM and the novel findings of [162] for the ADC.

Given that the ADC model differs from the ADM only by a few extra terms in the right-hand side of the correction step, and having the full analysis of the ADM in [166, 77] (including the energy equality, scale-similarity discussion and the microscale computation), one would think that similar results for the ADC are almost immediate. However, herein lies the issue: if one considers the difference between the ADM and the ADC to be the "extra" terms in the right-hand side, then one arrives at exactly the same energy equality for the ADC, and the rest of the discussion gets derailed.

The author of [162] took a different route, comparing the energy, energy dissipation rate and the power input of the ADC to the corresponding quantities from the *filtered* NSE. Working with $\bar{u}$ instead of u is justified, because both the ADM and the ADC aim to approximate the solution pair $(\bar{u}, p)$.

8.1 ENERGY BALANCE

Following [77], we derive the energy equality for the ADM by choosing the test function $v^h = w_1$ in (7.3) and integrating over time. This leads to

$$E_{ADM}(t) + \int_0^t \mathcal{E}_{ADM}(\tau)d\tau = E_{ADM}(0) + \int_0^t P_{ADM}(\tau)d\tau, \qquad (8.1)$$

where

DOI: 10.1201/9781003355977-8

$$E_{ADM}(t) = \text{model's energy} = \frac{1}{2}\left(\|w_1(t)\|^2 + \delta^2\|\nabla w_1(t)\|^2\right), \tag{8.2}$$

$$\mathcal{E}_{ADM}(t) = \text{model's dissipation rate} = \nu\left(\|\nabla w_1(t)\|^2 + \delta^2\|\Delta w_1(t)\|^2\right) \tag{8.3}$$

$$P_{ADM}(t) = \text{model's power input} = (f(t), w_1(t)). \tag{8.4}$$

Compared to the NSE, the ADM energy and energy dissipation rate have extra terms that are approaching zero quadratically fast as the filtering width δ is being reduced: these are $\delta^2\|\nabla w_1(t)\|^2$ and $\nu\delta^2\|\Delta w_1(t)\|^2$, respectively.

The discussion of the energy equality for the ADC starts with the NSE, written in terms of the filtered velocity $\bar{u}$. Using $u = \bar{u} - \delta^2\Delta\bar{u}$, we seek $(\bar{u}, \zeta, p) \in (X^h, X^h, Q^h)$ such that for any $(v^h, \xi^h, \chi^h) \in (X^h, X^h, Q^h)$

$$\begin{aligned}(\bar{u}_t, v^h) + \delta^2(\nabla\bar{u}_t, \nabla v^h) + \nu(\nabla\bar{u}, \nabla v^h) + \nu\delta^2(\nabla\zeta, \nabla v^h)\\ +b^*(\bar{u}, \bar{u}; v^h) - (p, \nabla\cdot v^h) = (f, v^h)\\ -\delta^2 b^*(\zeta, \bar{u}; v^h) - \delta^2 b^*(\bar{u}, \zeta; v^h) - \delta^4 b^*(\zeta, \zeta; v^h),\\ (\nabla\bar{u}, \nabla\xi^h) = (\zeta, \xi^h),\\ (\nabla\cdot\bar{u}, \chi^h) = 0.\end{aligned} \tag{8.5}$$

The energy equality for the filtered velocity $\bar{u}$ can be obtained by taking $v^h = \bar{u} - \delta^2\Delta\bar{u}$ in (8.5). It reads:

$$E_{\bar{u}}(t) + \int_0^t \mathcal{E}_{\bar{u}}(\tau)d\tau = E_{\bar{u}}(0) + \int_0^t P_{\bar{u}}(\tau)d\tau, \tag{8.6}$$

where

$$E_{\bar{u}}(t) = \frac{1}{2}\left(\|\bar{u}(t)\|^2 + 2\delta^2\|\nabla\bar{u}\|^2 + \delta^4\|\Delta\bar{u}\|^2\right), \tag{8.7}$$

$$\mathcal{E}_{\bar{u}}(t) = \nu\left(\|\nabla\bar{u}(t)\|^2 + 2\delta^2\|\Delta\bar{u}\|^2 + \delta^4\|\nabla(\Delta\bar{u})\|^2\right), \tag{8.8}$$

$$P_{\bar{u}}(t) = (f(t), \bar{u} - \delta^2\Delta\bar{u}). \tag{8.9}$$

Notice that, when compared to the quantities from (8.7)-(8.9), the energy and the energy dissipation rate of the ADM no longer have extraneous terms. In fact, there are some terms in (8.7)–(8.9) that don't have analogues in (8.2)–(8.4).

The authors of [162] noticed the structure of the power input in (8.9); they chose $v^h = -\delta^2\Delta w_1$ in (7.3), took $v^h = w_2$ in (7.4), added the two momentum equations and integrated over time. This led to the novel form of the energy balance equation with a numerical power input (wording introduced in [162]):

$$E_{ADC}(t) + \int_0^t \mathcal{E}_{ADC}(\tau)d\tau = E_{ADC}(0) + \int_0^t P_{ADC}(\tau)d\tau$$

$$+\int_0^t P_{\mathcal{N},ADC}(\tau)d\tau, \tag{8.10}$$

where the ADC quantities are comparable to those of the filtered NSE:

$$\begin{aligned}
&E_{ADC}(t) = \text{model's energy}\\
&= \frac{1}{2}\left(\|w_2(t)\|^2 + \delta^2\|\nabla w_2(t)\|^2 + \delta^2\|\nabla w_1(t)\|^2 + \delta^4\|\Delta w_1(t)\|^2\right), &(8.11)\\
&\mathcal{E}_{ADC}(t) = \text{model's dissipation rate}\\
&= \nu\left(\|\nabla w_2(t)\|^2 + \delta^2\|\Delta w_2(t)\|^2 + \delta^2\|\Delta w_1(t)\|^2 + \delta^4\|\nabla(\Delta w_1(t))\|^2\right), &(8.12)\\
&P_{ADC}(t) = \text{model's power input} = (f(t), w_2(t) - \delta^2\Delta w_1(t)), &(8.13)\\
&P_{\mathcal{N},ADC}(t) = \text{model's numerical power input} = (\mathcal{A}, \mathcal{B}), \text{ with}\\
&\mathcal{A} = w_1(t)\cdot\nabla w_1(t) - (w_1(t) - \delta^2\Delta w_1(t))\cdot\nabla(w_1(t) - \delta^2\Delta w_1(t)),\\
&\mathcal{B} = w_1(t) - \delta^2\Delta w_1(t) - w_2(t). &(8.14)
\end{aligned}$$

It is worth noting that, if the regularity assumptions of Theorem 18 are satisfied, both $\mathcal{A}$ and $\mathcal{B}$ are of the order $O(\delta^2)$, so when the filtering width δ is reduced, the numerical power input decays as $O(\delta^4)$.

8.2 KINETIC ENERGY SPECTRUM OF THE ADM AND THE ADC

Having the energy equality, one naturally proceeds to discuss the energy cascade(s). To do so, we take $\nu = 0$, $f = 0$. With the filtering width small enough, we argue that the numerical power input $P_{\mathcal{N},ADC}(t) = O(\delta^4)$ is negligible. Using the kinetic energy distribution functions, [77, 162] and following Kraichnan's dynamic argument, [163], leads to the following energy cascades for the ADM and the ADC.

$$\begin{aligned}
E_{ADM}(k) &\simeq \mathcal{E}_{ADM}^{2/3} k^{-5/3}, \text{ for } k \le \frac{1}{\delta}, &(8.15)\\
E_{ADM}(k) &\simeq \mathcal{E}_{ADM}^{2/3}\delta^{-2} k^{-11/3}, \text{ for } k \ge \frac{1}{\delta}.
\end{aligned}$$

$$\begin{aligned}
E_{ADC}(k) &\simeq \mathcal{E}_{ADC}^{2/3} k^{-5/3}, \text{ for } k \le \frac{1}{\delta}, &(8.16)\\
E_{ADC}(k) &\simeq \mathcal{E}_{ADC}^{2/3}\delta^{-4} k^{-17/3}, \text{ for } k \ge \frac{1}{\delta}.
\end{aligned}$$

Notice that both models correctly predict the energy cascade of the NSE in the inertial range $k \le \frac{1}{\delta}$. The models differ in the secondary cascades, which gives better insight into the action of the ADC. As is evident from the plot below, the ADC dissipates the subgrid scale energy more efficiently, than the ADM.

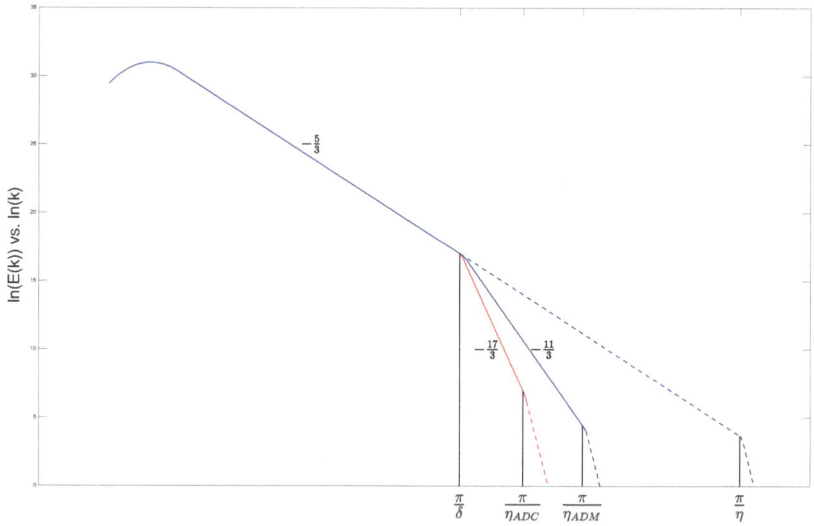

Figure 8.1 Kinetic energy spectrum.

The ADC microscale η_{ADC} will be discussed next, along with its positioning relative to the filtering width δ, the ADM microscale η_{ADM} and the NSE microscale η.

8.3 THE MICROSCALE OF THE ADC

The microscale discussion for the ADM [77] and the ADC [162] stems from two key ideas; firstly, at the smallest scales the model's Reynolds number should be of order $Re_{small} = O(1)$: dissipation becomes comparable to convection. Secondly, at statistical equilibrium the energy input at large scales should match the energy dissipation at the microscale.

The derivation process for the ADC microscale η_{ADC} starts in [162] with the expressions for the Reynolds numbers at the largest and smallest scales.

$$\begin{aligned} Re_{ADC,large} &= \frac{UL}{\nu\left(1+(\frac{\delta}{L})^2\right)^2}, \\ Re_{ADC,small} &= \frac{w_{small,ADC}\,\eta}{\nu\left(1+(\frac{\delta}{\eta})^2\right)^2}. \end{aligned} \tag{8.17}$$

The model's small-scale velocity can now be derived from $Re_{ADC,small} =$

$O(1)$. This velocity $w_{small,ADC}$ is then used in the formulation for the energy dissipation rate at small scales. The authors of [162] argue that two out of four terms in (8.12) are larger than the other two, and the small-scale energy dissipation rate is approximately equal to

$$\mathcal{E}_{small,ADC} \simeq \nu \left(\frac{\delta^2 w_{small,ADC}^2}{\eta^4} + \frac{\delta^4 w_{small,ADM}^2}{\eta^6} \right). \tag{8.18}$$

The small-scale ADM velocity $w_{small,ADM}$, used in (8.18), depends upon the ADM microscale η_{ADM}. Expressions for both of these quantities were derived in [77]. In particular, the ADM microscale was found to be

$$\eta_{ADM} \simeq Re^{-3/10} L^{2/5} \delta^{3/5}. \tag{8.19}$$

Using these ADM results and matching the rate of energy input at the ADC larger scales to the energy dissipation rate at the ADC smaller scales leads to the derivation of the ADC microscale, [162]:

$$\eta_{ADC} \simeq Re^{-1/5} L^{4/15} \delta^{11/15}. \tag{8.20}$$

Comparing the microscales of the ADM and the ADC, we immediately notice that the ADC truncates the small scales faster than the ADM, $\eta_{ADC} > \eta_{ADM}$. This is true, provided that the filtering width δ is chosen to be larger than the NSE microscale – a very natural assumption that is always true in modeling of turbulent flows. A more precise placement of the microscales gives

$$\eta_{ADC}^{3/2} \simeq \delta^{1/2} \eta_{ADM} \Leftrightarrow \eta_{ADC} \simeq \delta^{1/3} \eta_{ADM}^{2/3}. \tag{8.21}$$

The key takeaway message is the treatment of scales smaller than the cut-off scale $\frac{\pi}{\delta}$. Above the cut-off both the ADM and the ADC have the energy cascade that agrees with the NSE cascade in the inertial range. Below the cut-off length scale the ADC dissipates energy at a higher rate than the ADM, which further explains the increased efficiency of the ADC.

CHAPTER 9

Defect-Deferred Correction

The Full ADC Model for the NSE

9.1 FULL ADC WITH REDUCED TEMPORAL DISCRETIZATION ERROR: FREE LUNCH!

Let us investigate one possible improvement of the ADC turbulence model from Chapter 7. Now that we have introduced a correction step to reduce the modeling component of the overall error, why not use it to also reduce the temporal discretization error? The famous statement "there is no free lunch" does not apply here!

Indeed, if we consider a two-step (one defect and one correction) Algorithm 7.1.1 to build the ADC from the ADM, and use the Backward Euler method for time discretization, then the overall accuracy of the models will be of order $O(h^m + \delta^2 + \Delta t)$ and $O(h^m + \delta^4 + \Delta t)$ for the ADM and the ADC, respectively. At the same time, a deferred correction method (e.g., the one introduced for the ADM in Chapter 5) could increase the accuracy of the time discretization in the correction step solution – if built properly into Algorithm 7.1.1. This is what we will discuss in this chapter, following the findings in [168].

Let us start with the semidiscrete version of the ADM/ADC models (continuous in time). The ADM seeks $(w_1, \zeta_1, q_1) \in (X^h, X^h, Q^h)$ such that for any $(v^h, \xi^h, \chi^h) \in (X^h, X^h, Q^h)$

$$
\begin{aligned}
(w_{1,t}, v^h) + \delta^2(\nabla w_{1,t}, \nabla v^h) + \nu(\nabla w_1, \nabla v^h) + \nu\delta^2(\nabla \zeta_1, \nabla v^h) \\
+ b^*(w_1, w_1; v^h) - (q_1, \nabla \cdot v^h) = (f, v^h), \\
(\nabla w_1, \nabla \xi^h) = (\zeta_1, \xi^h), \\
(\nabla \cdot w_1, \chi^h) = 0,
\end{aligned}
\tag{9.1}
$$

DOI: 10.1201/9781003355977-9

subject to $w_1(0,x) = \overline{u}_0^\delta(x)$ and boundary conditions.

The correction step of the ADC reads: find $(w_2, \zeta_2, q_2) \in (X^h, X^h, Q^h)$ such that for any $(v^h, \xi^h, \chi^h) \in (X^h, X^h, Q^h)$

$$\begin{aligned}(w_{2,t}, v^h) + \delta^2(\nabla w_{2,t}, \nabla v^h) + \nu(\nabla w_2, \nabla v^h) + \nu\delta^2(\nabla \zeta_2, \nabla v^h) \\ + b^*(w_2, w_2; v^h) - (q_2, \nabla \cdot v^h) = (f, v^h) \\ - \delta^2 b^*(\zeta_1, w_1; v^h) - \delta^2 b^*(w_1, \zeta_1; v^h) - \delta^4 b^*(\zeta_1, \zeta_1; v^h), \\ (\nabla w_2, \nabla \xi^h) = (\zeta_2, \xi^h), \\ (\nabla \cdot w_2, \chi^h) = 0.\end{aligned} \tag{9.2}$$

The deferred correction method will now be implemented on top of the LES-C model. When combining defect (Algorithm 7.1.1) and deferred correction, we always have a choice, which procedure to use first. In this book we only consider the case where deferred correction comes last.

The authors in [168] used the deferred correction approach of [36] and [132], that we have already seen in Chapter 5. It adds extra terms to the correction step of the ADC. If the Backward Euler method is used for time discretization, the "full" ADC model seeks $(w_1^{n+1}, \zeta_1^{n+1}, q_1^{n+1})$, $(w_2^{n+1}, \zeta_2^{n+1}, q_2^{n+1}) \in (X^h, X^h, Q^h), \forall (v^h, \xi^h, \chi^h) \in (X^h, X^h, Q^h)$ at $t = t_{n+1}$, $n \geq 0$, with $\Delta t := t_{i+1} - t_i$, satisfying

$$\begin{aligned}(\frac{w_1^{n+1} - w_1^n}{\Delta t}, v^h) + \delta^2(\frac{\nabla w_1^{n+1} - \nabla w_1^n}{\Delta t}, \nabla v^h) + \nu(\nabla w_1^{n+1}, \nabla v^h) \\ + \nu\delta^2(\nabla \zeta_1^{n+1}, \nabla v^h) + b^*(w_1^{n+1}, w_1^{n+1}; v^h) - (q_1^{n+1}, \nabla \cdot v^h) \\ = (f(t_{n+1}), v^h), (\nabla w_1^{n+1}, \nabla \xi^h) = (\zeta_1^{n+1}, \xi^h), \\ (\nabla \cdot w_1^{n+1}, \chi^h) = 0.\end{aligned} \tag{9.3}$$

$$\begin{aligned}(\frac{w_2^{n+1} - w_2^n}{\Delta t}, v^h) + \delta^2(\frac{\nabla w_2^{n+1} - \nabla w_2^n}{\Delta t}, \nabla v^h) + \nu(\nabla w_2^{n+1}, \nabla v^h) \\ + \nu\delta^2(\nabla \zeta_2^{n+1}, \nabla v^h) + b^*(w_2^{n+1}, w_2^{n+1}; v^h) \\ - (q_2^{n+1}, \nabla \cdot v^h) = (\frac{f(t_{n+1}) + f(t_n)}{2}, v^h) \\ + \frac{\Delta t}{2}\nu(\frac{\nabla w_1^{n+1} - \nabla w_1^n}{\Delta t}, \nabla v^h) + \frac{\Delta t}{2}\nu\delta^2(\frac{\nabla \zeta_1^{n+1} - \nabla \zeta_1^n}{\Delta t}, \nabla v^h) \\ + \frac{1}{2}b^*(w_1^{n+1}, w_1^{n+1}; v^h) - \frac{1}{2}b^*(w_1^n, w_1^n; v^h) - \frac{\Delta t}{2}(\frac{q_1^{n+1} - q_1^n}{\Delta t}, \nabla \cdot v^h) \\ - \delta^2 b^*(\zeta_1^{n+1}, w_1^{n+1}; v^h) - \delta^2 b^*(w_1^{n+1}, \zeta_1^{n+1}; v^h) - \delta^4 b^*(\zeta_1^{n+1}, \zeta_1^{n+1}; v^h), \\ (\nabla w_2^{n+1}, \nabla \xi^h) = (\zeta_2^{n+1}, \xi^h), \\ (\nabla \cdot w_2^{n+1}, \chi^h) = 0.\end{aligned} \tag{9.4}$$

9.2 INCREASED ACCURACY OF THE FULL ADC – REASONING AND THEOREMS

Compared to the ADC formulation (7.6) from Chapter 7, there are five more terms in the right-hand side of (9.4) that ensure the increased temporal accuracy of this new model. This is done with essentially no extra cost, compared to the ADC, because the only modification concerns the change in the right-hand side of the correction step, with the known quantity. The claim of the increased temporal accuracy of the full ADC model (9.3)–(9.4) will now be investigated theoretically and numerically, following [168]. Needless to say, we will appreciate the earlier results on the stability and accuracy of the ADM, along with the accuracy proof of the gradient and the time derivative of the error. In fact, between the theorems of Chapter 5 and Chapter 7, we have all the necessary ingredients to prove the stability and increased accuracy of the full ADC. The operator norm $\|\cdot\|_A$, used in the theorems below, was defined in Chapter 7: $\|\phi\|_A^2 = \|\phi\|^2 + \delta^2\|\nabla\phi\|^2$.

Theorem 21 (Stability of the full ADC model). *Let w_2 satisfy (9.4) and let the assumptions of Theorems 11 and 16 be satisfied. Then, for $n = 0, \dots, N-1$*

$$\|w_2^{n+1}\|_A^2 + \nu\Delta t\sum_{i=0}^{n}\|\nabla w_2^{i+1}\|^2 + \delta^2\nu\Delta t\sum_{i=0}^{n}\|\zeta_2^{i+1}\|^2 \tag{9.5}$$
$$\leq \frac{C}{\nu^2}\left(\|w_2^0\|_A^2 + \frac{\Delta t}{\nu}\sum_{i=0}^{n}\left\|\frac{f(t_{i+1}) + f(t_i)}{2}\right\|_{-1}^2\right)$$

Theorem 22 (Accuracy of the Second Approximation). *Let the assumptions of Theorems 11 and 16 be satisfied. Then there exists a constant C such that the following accuracy bound holds for the solution (w_2, ζ_2, q_2) of (9.4):*

$$\|\bar{u}(t_{n+1}) - w_2^{n+1}\|_A^2 + \nu\Delta t\sum_{i+0}^{n}\|\nabla(\bar{u}(t_{n+1}) - w_2^{n+1})\|^2$$
$$+\nu\delta^2\Delta t\sum_{i=0}^{n}\|\zeta(t_{n+1}) - \zeta_2^{n+1}\|^2 \leq C(\Delta t^4 + \Delta t^2\delta^4 + \delta^8$$
$$+\Delta t\sum_{i=0}^{n}((1+\delta^2h^{-2})\inf_{v\in X^h}\|\nabla(\bar{u}(t_i) - v^i)\|^2 + (1+\delta^4h^{-2})\inf_{v\in X^h}\|\bar{u}_t(t_i) - v^i\|^2$$
$$+\inf_{q\in Q^h}\|p(t_i) - q^i\|^2 + \inf_{r\in X^h}\|(\zeta(t_i) - r^i)\|^2))$$

$$+\Delta t^2(\nu\Delta t\sum_{i=0}^{n}\inf_{v\in V^h}\left\|\frac{\nabla((\bar{u}(t_{i+1}) - v^{i+1}) - \bar{u}(t_i) - v^i))}{\Delta t}\right\|^2$$
$$+\Delta t\sum_{i=0}^{n}\inf_{q\in Q^h}\left\|\frac{(p(t_{i+1}) - q^{i+1}) - p(t_i) - q^i)}{\Delta t}\right\|^2$$

$$+\nu\delta^4\Delta t\sum_{i=0}^{n}\inf_{r\in V^h}\left\|\frac{(\zeta(t_{i+1})-r^{i+1})-\zeta(t_i)-r^i)}{\Delta t}\right\|^2))$$

9.3 NUMERICS

9.3.1 Convergence Rates at Different Values of the Reynolds Number: Asymptotic Convergence

The full ADC model (9.3)–(9.4) aims to reduced two out of three components of the overall error: the modeling error and the time discretization error. The authors of [168] introduced the following test to check the claimed convergence rates. The considered a flow in $\{0.5 \le x \le 1, 0.5 \le y \le 1, 0.25 \le t \le 0.75\}$, at $\nu = 10^{-3}$ and $\nu = 10^{-5}$, with a manufactured solution

$$\begin{aligned}
u_1 &= \tfrac{3}{4} + \tfrac{1}{4}\ \cos(2\pi(\mathrm{x}-\mathrm{t}))\sin(2\pi(y-t))e^{(-8\pi^2 t\nu)},\\
u_2 &= \tfrac{3}{4} - \tfrac{1}{4}\ \sin(2\pi(\mathrm{x}-\mathrm{t}))\cos(2\pi(y-t))e^{(-8\pi^2 t\nu)},\\
p &= -\tfrac{1}{64}\ \cos(4\pi(\mathrm{x}-\mathrm{t}))\cos(4\pi(y-t))e^{(-16\pi^2 t\nu)}.
\end{aligned}$$

The errors and convergence rates were computed at a fixed spatial mesh with the mesh diameter $h = 1/128$. Both the filtering width δ and the time step Δt were reduced (halved); based on the theoretical results, we expect first order accuracy from the ADM and second order accuracy from the ADC. In fact, we expect to see $O(\Delta t + \delta^2)$ and $O((\Delta t)^2 + \delta^4)$, respectively, but this chapter focuses more on the increased temporal accuracy of the full ADC.

There are other tests presented in [168], including the computation of drag and lift coefficients and the 2-D flow past a step, but we feel that some important conclusions about the performance of the ADC can already be drawn from Tables 9.1–9.2. Of course, one has to keep in mind that this test problem has nothing to do with a really turbulent setting; indeed, no matter how small the viscosity coefficient ν, the approximated manufactured solution is still a smooth function. Still, we can make connections between

TABLE 9.1 Errors in (9.3) and (9.4), $L^2(0,T;L^2(\Omega))$ norm, $\nu = 10^{-3}$, spatial mesh is fixed at $h = \frac{1}{128}$, while the time step and filtering width are refined

$\delta = \Delta t$	$\|\|\bar{u} - w_1\|\|_{L^2(0,T;L^2(\Omega))}$	$\|\|\bar{u} - w_2\|\|_{L^2(0,T;L^2(\Omega))}$	ADM rates	ADC rates
h	0.00227971	0.000657181	-	-
h/2	0.00117971	0.000147313	0.95	2.16
h/4	0.000600011	2.97E-05	0.98	2.31
h/8	0.000302609	5.58E-06	0.99	2.41

TABLE 9.2 Errors in (9.3) and (9.4), $L^2(0,T;H^1(\Omega))$ norm, $\nu = 10^{-3}$, spatial mesh is fixed at $h = \frac{1}{128}$, while the time step and filtering width are refined

$\delta = \Delta t$	$\|\|\bar{u} - w_1\|\|_{L^2(0,T;L^2(\Omega))}$	$\|\|\bar{u} - w_2\|\|_{L^2(0,T;L^2(\Omega))}$	ADM rates	ADC rates
h	0.0900065	0.0940873	-	-
h/2	0.0544724	0.0311926	0.72	1.59
h/4	0.0311129	0.00846719	0.81	1.88
h/8	0.0169704	0.00188764	0.87	2.17

the traditional expectations of a predictor–corrector scheme and the results obtained by the ADC.

We notice both the higher convergence rates and the smaller sizes of the overall error of the ADC, compared to the ADM. We also notice that the size of the ADC error in the next-to-last refinement of $\delta, \Delta t$ is less than the ADM error in the case of the smallest values of $\delta, \Delta t$. This correlates well with the concept of "accuracy per computational cost" from Chapter 7: it would take less computational resources and CPU time for the ADC solution to get within the prescribed tolerance from the reference solution, because it can be accomplished in half the number of time steps and with a larger value of δ.

We also notice the asymptotic convergence, typical of the defect correction methods: it takes several refinements of the problem parameters, in order to see the claimed convergence rates. It is definitely of interest to check, how this is affected by the increased Reynolds number, so the authors of [168] repeated the convergence rates computation with $\nu = 10^{-5}$. The results are presented in Tables 9.3–9.4 below.

Again, we notice that the ADC with larger values of $\delta, \Delta t$ can produce better results than the ADM with smaller $\delta, \Delta t$. The convergence rates of the ADC increasing past the expected second order indicate that in this test problem the temporal discretization component of the overall error is less pronounced than the modeling error, which in case of the ADC is claimed to be fourth order of smallness, $O(\delta^4)$.

9.3.2 Flow Past a 3-D Step

We want to present one more test from [168]. In it, the flow enters the left wall of the 3-D domain $\Omega = [0, 40] \times [0, 10] \times [0, 10]$, with the step at $5 \le x \le 6, 0 \le z \le 1$. At the Reynolds number $Re = 600$ we expect to see a wave front forming past the step and then traveling to the right. The reference solution, computed by direct numerical simulation (DNS, no modeling) at the spatial mesh with

33, 204 degrees of freedom, is presented below. Time discretization was done with the Crank–Nicolson method (second order accurate) at $\Delta t = 1/16$. This is to be compared with the ADM and ADC solutions on the same spatial mesh, with $\delta = 1/8$, $\Delta t = 1/8$.

Figure 9.1 shows that the ADM fails to produce the traveling wave front, whereas both the reference solution and the ADC solution seem to capture this wave front traveling near the plane $x = 17$.

9.4 SHOULD WE ALWAYS USE THE FULL VERSION OF THE ADC?

Despite the fact that increased temporal accuracy is obtained by the deferred correction "for free", many would argue that the combination of the ADC model of Chapter 7 with a higher accurate time discretization, used in place of the Backward Euler method, would create a more efficient model than the full ADC proposed here. Indeed, there is an abundance of time discretization techniques, and one can hope to find a good fit for the problem at hand – in terms of stability properties and the desired order of accuracy. However, in some situations we want to be able to "build" on top of an existing method that is low order accurate, but fits uniquely the problem at hand. In particular, Chapters 10 and 11 discuss coupled Navier–Stokes systems, where monolithic methods are too expensive to use if the flow is turbulent. The MHD system in Chapter 10 and the fluid–fluid interaction problem in Chapter 11 are to be decoupled before a turbulence model is built. Stable partitioning (decoupling) methods are hard to find; those that exist, are usually low order accurate. In such situations, where there are no alternative Runge–Kutta or multistep methods to use, improving the temporal accuracy via deferred correction could be of interest.

TABLE 9.3 Errors in (9.3) and (9.4), $L^2(0,T;L^2(\Omega))$ norm, $\nu = 10^{-5}$, spatial mesh is fixed at $h = \frac{1}{128}$, while the time step and filtering width are refined

$\delta = \Delta t$	$\|\|\bar{u} - w_1\|\|_{L^2(0,T;L^2(\Omega))}$	$\|\|\bar{u} - w_2\|\|_{L^2(0,T;L^2(\Omega))}$	ADM rates	ADC rates
h	0.00245446	0.00981675	-	-
h/2	0.00127319	0.00290769	0.95	1.76
h/4	0.00066767	0.00107661	0.93	1.43
h/8	0.000381481	0.000369831	0.81	1.54
h/16	0.00022466	8.21E-05	0.76	2.17
h/32	0.000121875	1.32E-05	0.88	2.64

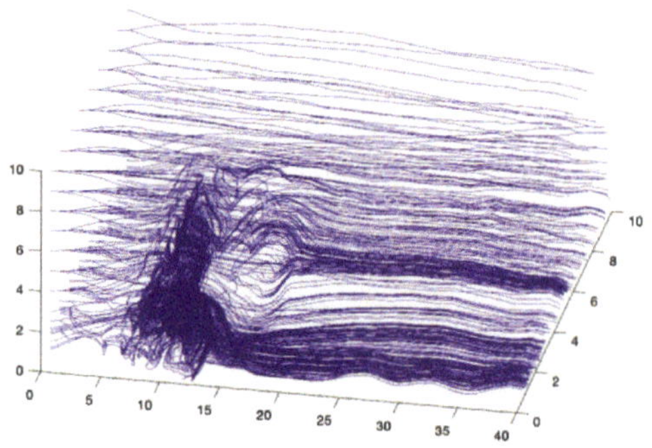

(a)

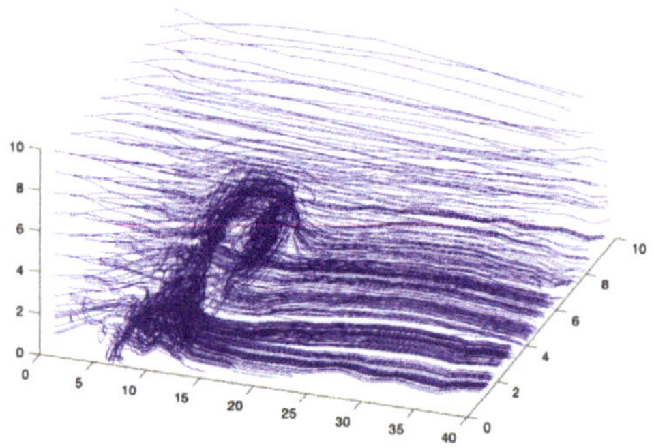

(b)

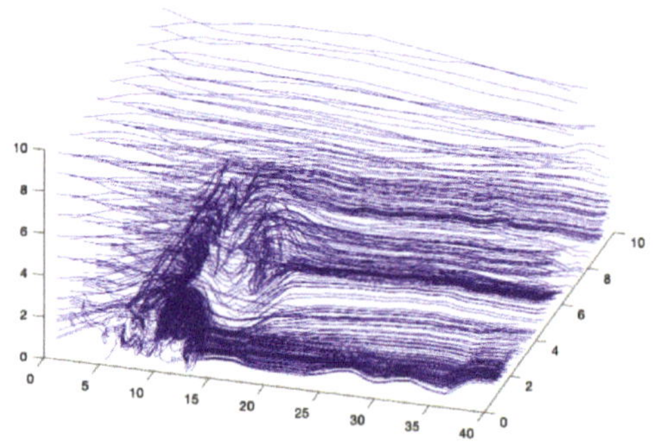

(c)

Figure 9.1 Solutions at a coarse mesh. (a) Reference CN-NSE solution at $\Delta t = 1/16$. (b) ADC defect step solution at $\Delta t = 1/8$. (c) ADC correction step solution at $\Delta t = 1/8$.

TABLE 9.4 Errors in (9.3) and (9.4), $L^2(0,T;H^1(\Omega))$ norm, $\nu = 10^{-5}$, spatial mesh is fixed at $h = \frac{1}{128}$, while the time step and filtering width are refined

			ADM	ADC
$\delta = \Delta t$	$\|\|\bar{u} - w_1\|\|_{L^2(0,T;L^2(\Omega))}$	$\|\|\bar{u} - w_2\|\|_{L^2(0,T;L^2(\Omega))}$	rates	rates
h	0.223553	5.73065	-	-
h/2	0.170311	2.70714	0.39	1.08
h/4	0.16385	0.959415	0.06	1.50
h/8	0.163266	0.291716	0.01	1.72
h/16	0.119295	0.0609031	0.45	2.26
h/32	0.0683414	0.00941398	0.80	2.69

CHAPTER 10

The Full ADC Model for Magnetohydrodynamics

In this chapter we return to the MagnetoHydroDynamic (MHD) flows discussed in Chapter 3. We will consider the case of a MHD flow at high Reynolds and magnetic Reynolds numbers, and test the full ADC model of Chapter 9 in the MHD setting. The need for the deferred correction stems from the desire to decouple the MHD system, in order to make the computations more efficient. Just like in Chapter 3, the decoupling method of [73] is utilized; this IMEX scheme is stable, but only first order accurate. The deferred correction technique, introduced in Chapter 3, is now a part of the full ADC approach of Chapter 9. Thus, following [164], we aim to build the full ADC for the MHD.

The MHD system will be written in Elsässer variables, in order to use the IMEX decoupling method of [73]. First, the magnetic field is written as the sum of mean and fluctuations, $B = B_0 + b$. Next, we switch from the velocity u and magnetic fluctuations b to the Elsässer variables $z_+ = u+b$, $z_- = u-b$. The Elsässer formulation is well-known and has been studied by many researchers, see, e.g., [173, 174, 175, 176].

The decoupled model for $z^i_{+/-} \approx z_{+/-}(t_i)$, introduced and proven to be stable and first-order accurate in time in [73], reads:

$$
\begin{aligned}
\frac{z_+^{n+1} - z_+^n}{\Delta t} - B_0 \cdot \nabla z_+^{n+1} + z_-^n \cdot \nabla z_+^{n+1} - \frac{\nu + \nu_m}{2} \Delta z_+^{n+1} \\
- \frac{\nu - \nu_m}{2} \Delta z_-^n + \nabla p_+^{n+1} = f_+(t_{n+1}), \\
\nabla \cdot z_+^{n+1} = 0, \\
\frac{z_-^{n+1} - z_-^n}{\Delta t} + B_0 \cdot \nabla z_-^{n+1} + z_+^n \cdot \nabla z_-^{n+1} - \frac{\nu + \nu_m}{2} \Delta z_-^{n+1} \\
- \frac{\nu - \nu_m}{2} \Delta z_+^n + \nabla p_-^{n+1} = f_-(t_{n+1}), \\
\nabla \cdot z_-^{n+1} = 0.
\end{aligned}
\tag{10.1}
$$

DOI: 10.1201/9781003355977-10

The mean magnetic field plays an important role in MHD turbulence, for example, it can make the turbulence anisotropic; suppress the turbulence by decreasing energy cascade, etc. In the presence of a strong mean magnetic field, $\boldsymbol{z}^{+}$ and $\boldsymbol{z}^{-}$ wavepackets travel in opposite directions with the phase velocity of $\boldsymbol{B}_{\circ}$, and interact weakly. For Kolmogorov's and Iroshnikov/Kraichnan's phenomenological theories of MHD isotropic and anisotropic turbulence, see [60, 62, 45, 67, 74, 70, 52, 55, 75].

10.1 THE IMEX-BASED ADM FOR DECOUPLED MHD IN ELSÄSSER VARIABLES

In order to derive the full ADC model, first we must obtain the ADM for Elsässer MHD. The ADM for the MHD was presented in [85]; in the Elsässer MHD formulation, the ADM seeks to approximate the filtered Elsässer variables and non-filtered pressure fields $(\bar{z}_{+}, \bar{z}_{-}, p_{+}, p_{-})$ with the model solutions $(w_{+}, w_{-}, q_{+}, q_{-}) \in (X^h, X^h, Q^h, Q^h)$. The model, [164], introduces different filtering widths $\delta_{+/-}$ for extra flexibility:

$$
\begin{aligned}
\left(\frac{w_{+}^{n+1}-w_{+}^{n}}{\Delta t}, v\right) + \delta_{+}^{2}\left(\nabla\left(\frac{w_{+}^{n+1}-w_{+}^{n}}{\Delta t}\right), \nabla v\right) \qquad (10.2)\\
-b^{*}(B_0, w_{+}^{n+1}; v) + b^{*}(w_{-}^{n}, w_{+}^{n+1}; v)\\
+\frac{\nu+\nu_m}{2}(\nabla w_{+}^{n+1}, \nabla v) + \frac{\nu+\nu_m}{2}\delta_{+}^{2}(\nabla \zeta_{+}^{n+1}, \nabla v) + \frac{\nu-\nu_m}{2}(\nabla w_{-}^{n}, \nabla v)\\
+\frac{\nu-\nu_m}{2}\delta_{-}^{2}(\nabla \zeta_{-}^{n}, \nabla v) - (q_{+}^{n+1}, \nabla\cdot v) = (f_{+}(t_{n+1}), v), \forall v \in X^h,\\
(\nabla\cdot w_{+}^{n+1}, q) = 0, \forall q \in Q^h,\\
(\nabla w_{+}^{n+1}, \nabla \xi) = (\zeta_{+}^{n+1}, \xi), \forall \xi \in X^h,\\
(\frac{w_{-}^{n+1}-w_{-}^{n}}{\Delta t}, v) + \delta_{-}^{2}\left(\nabla\left(\frac{w_{-}^{n+1}-w_{-}^{n}}{\Delta t}\right), \nabla v\right)\\
+b^{*}(B_0, w_{-}^{n+1}; v) + b^{*}(w_{+}^{n}, w_{-}^{n+1}; v)\\
+\frac{\nu+\nu_m}{2}(\nabla w_{-}^{n+1}, \nabla v) + \frac{\nu+\nu_m}{2}\delta_{-}^{2}(\nabla \zeta_{-}^{n+1}, \nabla v) + \frac{\nu-\nu_m}{2}(\nabla w_{+}^{n}, \nabla v)\\
+\frac{\nu-\nu_m}{2}\delta_{+}^{2}(\nabla \zeta_{+}^{n}, \nabla v) - (q_{-}^{n+1}, \nabla\cdot v) = (f_{-}(t_{n+1}), v), \forall v \in X^h,\\
(\nabla\cdot w_{-}^{n+1}, q) = 0, \forall q \in Q^h,\\
(\nabla w_{-}^{n+1}, \nabla \xi) = (\zeta_{-}^{n+1}, \xi), \forall \xi \in X^h.
\end{aligned}
$$

Comparing this ADM-MHD model with the IMEX MHD scheme (3.7)–(3.8) from Chapter 3, we notice the added terms with coefficient δ^2; they have the familiar structure of the corresponding terms from ADM-NSE, with the mixed variables $\zeta_{+/-}$ introduced to keep using C^0 finite element spaces.

10.2 FULL ADC FOR THE DECOUPLED MHD

Building the ADC via Algorithm 7.1.1 continues with the choice of the quantities of interest. Our models seek to approximate $(\bar{z}_+, \bar{z}_-, q_+, q_-)$, so first we must rewrite the Elsässer MHD in these variables. Given below is the mixed formulation of the Elsässer MHD, with quantities $\bar{z}_+ - \delta_+^2 \Delta \bar{z}_+$ and $\bar{z}_- - \delta_-^2 \Delta \bar{z}_-$ replacing z_+ and z_-.

$$\begin{aligned}
&(\bar{z}_{+,t}, v) + \delta_+^2(\nabla \bar{z}_{+,t}, \nabla v) - b^*(B_0, \bar{z}_+; v) - \delta_+^2 b^*(B_0, \tilde{\zeta}_+; v) \\
&+ b^*(\bar{z}_-, \bar{z}_+; v) + \delta_+^2 b^*(\bar{z}_-, \tilde{\zeta}_+; v) + \delta_-^2 b^*(\tilde{\zeta}_-, \bar{z}_+; v) + \delta_-^2 \delta_+^2 b^*(\tilde{\zeta}_-, \tilde{\zeta}_+; v) \\
&+ \frac{\nu + \nu_m}{2}(\nabla \bar{z}_+, \nabla v) + \frac{\nu + \nu_m}{2}\delta_+^2(\nabla \tilde{\zeta}_+, \nabla v) + \frac{\nu - \nu_m}{2}(\nabla \bar{z}_+, \nabla v) \\
&+ \frac{\nu - \nu_m}{2}\delta_-^2(\nabla \tilde{\zeta}_-, \nabla v) - (p_+, \nabla \cdot v) = (f_+, v), \forall v \in X, \\
&(\nabla \cdot \bar{z}_+, q) = 0, \forall q \in Q, \\
&(\nabla \bar{z}_+, \nabla \xi) = (\tilde{\zeta}_+, \xi), \forall \xi \in X. \\
&(\bar{z}_{-,t}, v) + \delta_-^2(\nabla \bar{z}_{-,t}, \nabla v) + b^*(B_0, \bar{z}_-; v) + \delta_-^2 b^*(B_0, \tilde{\zeta}_-; v) \\
&+ b^*(\bar{z}_+, \bar{z}_-; v) + \delta_-^2 b^*(\bar{z}_+, \tilde{\zeta}_-; v) + \delta_+^2 b^*(\tilde{\zeta}_+, \bar{z}_-; v) + \delta_+^2 \delta_-^2 b^*(\tilde{\zeta}_+, \tilde{\zeta}_-; v) \\
&+ \frac{\nu + \nu_m}{2}(\nabla \bar{z}_-, \nabla v) + \frac{\nu + \nu_m}{2}\delta_-^2(\nabla \tilde{\zeta}_-, \nabla v) + \frac{\nu - \nu_m}{2}(\nabla \bar{z}_-, \nabla v) \\
&+ \frac{\nu - \nu_m}{2}\delta_+^2(\nabla \tilde{\zeta}_+, \nabla v) - (p_-, \nabla \cdot v) = (f_-, v), \forall v \in X, \\
&(\nabla \cdot \bar{z}_-, q) = 0, \forall q \in Q, \\
&(\nabla \bar{z}_-, \nabla \xi) = (\tilde{\zeta}_-, \xi), \forall \xi \in X.
\end{aligned} \tag{10.3}$$

Compare (10.3) to (10.2): the red terms in (10.3) are the only ones that have no counterparts in (10.2). These will be used in constructing the ADC: similar terms, depending upon the defect step solution, will appear in the right-hand side of the correction equation.

The correction step of the full ADC, built on the defect step (10.2), will also incorporate the deferred correction procedure from Chapter 3, [76]. This full ADC-MHD model, introduced in [164], seeks to approximate the filtered velocities and non-filtered pressures $(\bar{z}_+, \bar{z}_-, p_+, p_-)$ with $(W_+, W_-, r_+, r_-) \in (X^h, X^h, Q^h, Q^h)$ via

$$\begin{aligned}
&\left(\frac{W_+^{n+1} - W_+^n}{\Delta t}, v\right) + \delta_+^2\left(\nabla\left(\frac{W_+^{n+1} - W_+^n}{\Delta t}\right), \nabla v\right) \\
&\quad - b^*(B_0, W_+^{n+1}; v) + b^*(W_-^n, W_+^{n+1}; v) \\
&\quad + \frac{\nu + \nu_m}{2}(\nabla W_+^{n+1}, \nabla v) + \frac{\nu + \nu_m}{2}\delta_+^2(\nabla Z_+^{n+1}, \nabla v) \\
&+ \frac{\nu - \nu_m}{2}(\nabla W_-^n, \nabla v) + \frac{\nu - \nu_m}{2}\delta_-^2(\nabla Z_-^n, \nabla v) - (r_+^{n+1}, \nabla \cdot v) \\
&= \left(\frac{f_+(t_{n+1}) + f_+(t_n)}{2}, v\right) - \frac{1}{2}b^*(B_0, w_+^{n+1}; v)
\end{aligned} \tag{10.4}$$

$$
\begin{aligned}
&+\frac{1}{2}b^*(B_0,w_+^n;v)+b^*(w_-^n,w_+^{n+1};v)-\frac{1}{2}b^*(w_-^{n+1},w_+^{n+1};v)-\frac{1}{2}b^*(w_-^n,w_+^n;v)\\
&+\frac{\nu+\nu_m}{4}\Delta t\left(\nabla\left(\frac{w_+^{n+1}-w_+^n}{\Delta t}\right),\nabla v\right)+\frac{\nu+\nu_m}{4}\delta_+^2\Delta t\left(\nabla\left(\frac{\zeta_+^{n+1}-\zeta_+^n}{\Delta t}\right),\nabla v\right)\\
&-\frac{\nu-\nu_m}{4}\Delta t\left(\nabla\left(\frac{w_-^{n+1}-w_-^n}{\Delta t}\right),\nabla v\right)-\frac{\nu-\nu_m}{4}\delta_-^2\Delta t\left(\nabla\left(\frac{\zeta_-^{n+1}-\zeta_-^n}{\Delta t}\right),\nabla v\right)\\
&-\frac{1}{2}\Delta t\left(\frac{q_+^{n+1}-q_+^n}{\Delta t},\nabla\cdot v\right)+\delta_+^2b^*\left(B_0,\frac{\zeta_+^{n+1}+\zeta_+^n}{2};v\right)-\frac{1}{2}\delta_+^2b^*(w_-^{n+1},\zeta_+^{n+1};v)\\
&\qquad-\frac{1}{2}\delta_+^2b^*(w_-^n,\zeta_+^n;v)-\frac{1}{2}\delta_-^2b^*(\zeta_-^{n+1},w_+^{n+1};v)-\frac{1}{2}\delta_-^2b^*(\zeta_-^n,w_+^n;v)\\
&\qquad-\frac{1}{2}\delta_-^2\delta_+^2b^*(\zeta_-^{n+1},\zeta_+^{n+1};v)-\frac{1}{2}\delta_-^2\delta_+^2b^*(\zeta_-^n,\zeta_+^n;v),\forall v\in X^h,\\
&(\nabla\cdot W_+^{n+1},q)=0,\forall q\in Q^h,\\
&(\nabla W_+^{n+1},\nabla\xi)=(Z_+^{n+1},\xi),\forall\xi\in X^h,\\
&\left(\frac{W_-^{n+1}-W_-^n}{\Delta t},v\right)+\delta_-^2\left(\nabla\left(\frac{W_-^{n+1}-W_-^n}{\Delta t}\right),\nabla v\right)\\
&\qquad+b^*(B_0,W_-^{n+1};v)+b^*(W_+^n,W_-^{n+1};v)\\
&+\frac{\nu+\nu_m}{2}(\nabla W_-^{n+1},\nabla v)+\frac{\nu+\nu_m}{2}\delta_-^2(\nabla Z_-^{n+1},\nabla v)+\frac{\nu-\nu_m}{2}(\nabla W_+^n,\nabla v)\\
&\qquad+\frac{\nu-\nu_m}{2}\delta_+^2(\nabla Z_+^n,\nabla v)-(r_-^{n+1},\nabla\cdot v)\\
&=\left(\frac{f_-(t_{n+1})+f_-(t_n)}{2},v\right)+\frac{1}{2}b^*(B_0,w_-^{n+1};v)\\
&-\frac{1}{2}b^*(B_0,w_-^n;v)+b^*(w_+^n,w_-^{n+1};v)-\frac{1}{2}b^*(w_+^{n+1},w_-^{n+1};v)-\frac{1}{2}b^*(w_+^n,w_-^n;v)\\
&+\frac{\nu+\nu_m}{4}\Delta t\left(\nabla\left(\frac{w_-^{n+1}-w_-^n}{\Delta t}\right),\nabla v\right)+\frac{\nu+\nu_m}{4}\delta_-^2\Delta t\left(\nabla\left(\frac{\zeta_-^{n+1}-\zeta_-^n}{\Delta t}\right),\nabla v\right)\\
&-\frac{\nu-\nu_m}{4}\Delta t\left(\nabla\left(\frac{w_+^{n+1}-w_+^n}{\Delta t}\right),\nabla v\right)-\frac{\nu-\nu_m}{4}\delta_+^2\Delta t\left(\nabla\left(\frac{\zeta_+^{n+1}-\zeta_+^n}{\Delta t}\right),\nabla v\right)\\
&-\frac{1}{2}\Delta t\left(\frac{q_-^{n+1}-q_-^n}{\Delta t},\nabla\cdot v\right)-\delta_-^2b^*\left(B_0,\frac{\zeta_-^{n+1}+\zeta_-^n}{2};v\right)-\frac{1}{2}\delta_-^2b^*(w_+^{n+1},\zeta_-^{n+1};v)\\
&\qquad-\frac{1}{2}\delta_-^2b^*(w_+^n,\zeta_-^n;v)-\frac{1}{2}\delta_+^2b^*(\zeta_+^{n+1},w_-^{n+1};v)-\frac{1}{2}\delta_+^2b^*(\zeta_+^n,w_-^n;v)\\
&\qquad-\frac{1}{2}\delta_+^2\delta_-^2b^*(\zeta_+^{n+1},\zeta_-^{n+1};v)-\frac{1}{2}\delta_+^2\delta_-^2b^*(\zeta_+^n,\zeta_-^n;v),\forall v\in X^h,\\
&(\nabla\cdot W_-^{n+1},q)=0,\forall q\in Q^h,\\
&(\nabla W_-^{n+1},\nabla\xi)=(Z_-^{n+1},\xi),\forall\xi\in X^h.
\end{aligned}
$$

The blue terms in (10.4) aim to increase the temporal accuracy from first to the second order via deferred correction. The red terms aim to reduce the modeling error of the ADM.

Note that, despite the plethora of extra terms in (10.4), the defect and correction steps of the model (10.2)–(10.4) are almost equal in computational complexity, as the left-hand sides of (10.2) and (10.4) have identical structure, and the right-hand side of (10.4) contains only the known quantities. This leads to efficient parallelization of the model, discussed in [164].

10.3 STABILITY OF THE FULL ADC-MHD MODEL

The results below make use of the following operator norms:

$$\|\phi\|^2_{A_{+/-}} = \|\phi\|^2 + \delta^2_{+/-}\|\nabla\phi\|^2.$$

The ADM-MHD model (10.2) must be investigated first, so we start with the stability and accuracy of the defect step approximations (w_+, w_-) – see the following two theorems from [164].

Theorem 23. *Let w_+, w_- satisfy (10.2). Let $f_+, f_- \in L^2(0,T;H^{-1}(\Omega))$. Then*

$$\begin{aligned} &\|w_+^{n+1}\|^2_{A_+} + \|w_-^{n+1}\|^2_{A_-} \\ &+ \min(\nu,\nu_m)\Delta t\sum_{i=1}^{n+1}(\|\nabla w_+^i\|^2 + \|\nabla w_-^i\|^2) \\ &+ 2\min(\nu,\nu_m)\Delta t\sum_{i=1}^{n+1}(\delta_+^2\|\zeta_+^i\|^2 + \delta_-^2\|\zeta_-^i\|^2) \\ \le\ &\|w_+^0\|^2_{A_+} + \|w_-^0\|^2_{A_-} + \frac{1}{\min(\nu,\nu_m)}\Delta t\sum_{i=1}^{n+1}(\|f_+(t_i)\|^2_{-1} + \|f_-(t_i)\|^2_{-1}). \end{aligned} \tag{10.5}$$

Theorem 24. *Given a final time $T > 0$ and time step $\Delta t > 0$, let $t_n = n\Delta t$ for $n = 0, 1, ..., N$. Let $\delta = max(\delta_+, \delta_-) \le C_0 h$ for some $C_0 = C_0(\nu, \nu_m)$. Then, provided the true solution satisfies the regularity assumptions*

$$\begin{aligned} \bar{z}_{+/-} &\in L^2((0,T);L^{(\infty)}(\Omega)), \\ \nabla\bar{z}_{+/-} &\in L^2((0,T);L^{(\infty)}(\Omega)) \\ \Delta\bar{z}_{+/-} &\in L^2((0,T);L^2(\Omega)), \\ \partial_t\bar{z}_{+/-} &\in L^\infty((0,T);L^{(\infty)}(\Omega)), \\ \partial_{tt}\bar{z}_{+/-} &\in L^2((0,T);L^2(\Omega)), \\ \partial_t\nabla\bar{z}_{+/-} &\in L^2((0,T);L^2(\Omega)) \\ \partial_{tt}\nabla\bar{z}_{+/-} &\in L^2((0,T);L^2(\Omega)), \end{aligned}$$

the following error estimate holds

$$\begin{aligned} &\left\|(\bar{z}_+(t_N) - \omega_+^N)\right\|^2_{A_+} + \left\|(\bar{z}_-(t_N) - \omega_-^N)\right\|^2_{A_-} \\ &+2\Delta t\sum_{n=0}^{N-1}\left\{\frac{\nu\nu_m}{(\nu+\nu_m)}[\left\|\nabla(\bar{z}_+(t_n) - \omega_+^n)\right\|^2 + \left\|\nabla(\bar{z}_-(t_n) - \omega_-^n)\right\|^2]\right. \end{aligned}$$

$$
\begin{aligned}
&+ \frac{(\nu+\nu_m)}{8}\delta^2(\|\tilde{\zeta}_+(t_n)-\zeta_+^n\|^2+\|\tilde{\zeta}_-(t_n)-\zeta_-^n\|^2)\Big\}\\
&\leq exp\Big(C_1\Delta t\sum_{n=0}^{N-1}[\|\nabla\bar{z}_+(t_{n+1})\|_\infty^2+\|\bar{z}_+(t_{n+1})\|_\infty^2\\
&\qquad +\|\nabla\bar{z}_-(t_{n+1})\|_\infty^2+\|\bar{z}_-(t_{n+1})\|_\infty^2\\
&+\frac{1}{(\nu+\nu_m)}\|\bar{z}_+(t_{n+1})\|_\infty^2+\frac{h^{-1}}{(\nu+\nu_m)}\inf_{v^n\in V^h}\|\nabla(\bar{z}_+(t_{n+1})-v^n)\|^2\\
&+\frac{1}{(\nu+\nu_m)}\|\bar{z}_-(t_{n+1})\|_\infty^2+\frac{h^{-1}}{(\nu+\nu_m)}\inf_{v^n\in V^h}\|\nabla(\bar{z}_-(t_{n+1})-v^n)\|^2]\Big)\\
&\times\Big\{C_2\Delta t\sum_{n=0}^{N-1}\Big(\frac{(\nu-\nu_m)^2}{(\nu+\nu_m)}\delta^4[\inf_{v^n\in V^h}\|\nabla(\tilde{\zeta}_+(t_n)-v^n)\|^2\\
&\qquad+\inf_{v^n\in V^h}\|\nabla(\tilde{\zeta}_-(t_n)-v^n)\|^2]\\
&+(\nu+\nu_m)\delta^2[\inf_{v^n\in V^h}\|\nabla(\tilde{\zeta}_+(t_{n+1})-v^n)\|^2+\inf_{v^n\in V^h}\|\nabla(\tilde{\zeta}_-(t_{n+1})-v^n)\|^2]\\
&\qquad+\inf_{v^n\in V^h}\|\nabla(\bar{z}_+(t_{n+1})-v^n)\|^2[(\nu+\nu_m)^2\\
&\qquad+\|\bar{z}_-(t_{n+1})\|_\infty^2+\frac{h^{-1}}{(\nu+\nu_m)}\inf_{v^n\in V^h}\|\nabla(\bar{z}_-(t_n)-v^n)\|^2]\\
&\qquad+\inf_{v^n\in V^h}\|\nabla(\bar{z}_-(t_{n+1})-v^n)\|^2[(\nu+\nu_m)^2\\
&\qquad+\|\bar{z}_+(t_{n+1})\|_\infty^2+\frac{h^{-1}}{(\nu+\nu_m)}\inf_{v^n\in V^h}\|\nabla(\bar{z}_+(t_n)-v^n)\|^2]\\
&+\frac{(\nu-\nu_m)^2}{(\nu+\nu_m)}\Big[\inf_{v^n\in V^h}\|\nabla(\bar{z}_+(t_n)-v^n)\|^2+\inf_{v^n\in V^h}\|\nabla(\bar{z}_-(t_n)-v^n)\|^2\Big]\\
&\qquad+\inf_{v^n\in V^h}\|\nabla(\bar{z}_+(t_{n+1})-v^n)\|^2\Big[\frac{1}{(\nu+\nu_m)}\|\bar{z}_-(t_{n+1})\|_\infty^2\Big]\\
&\qquad+\inf_{v^n\in V^h}\|\nabla(\bar{z}_-(t_{n+1})-v^n)\|^2\Big[\frac{1}{(\nu+\nu_m)}\|\bar{z}_+(t_{n+1})\|_\infty^2\Big]\\
&+\inf_{v^n\in V^h}\|\nabla(\bar{z}_+(t_n)-v^n)\|^2\Big[\frac{1}{(\nu+\nu_m)}\|\bar{z}_-(t_{n+1})\|_\infty^2+\|\nabla\bar{z}_-(t_{n+1})\|_\infty^2\Big]\\
&+\inf_{v^n\in V^h}\|\nabla(\bar{z}_-(t_n)-v^n)\|^2\Big[\frac{1}{(\nu+\nu_m)}\|\bar{z}_+(t_{n+1})\|_\infty^2+\|\nabla\bar{z}_+(t_{n+1})\|_\infty^2\Big]\\
&+\frac{\Delta t^2}{(\nu+\nu_m)}[\inf_{v^n\in V^h}\|(\bar{z}_{+,t}(t_{n+1})-v^n)\|^2+\inf_{v^n\in V^h}\|(\bar{z}_{-,t}(t_{n+1})-v^n)\|^2]\\
&+\frac{\Delta t^2\delta^4}{(\nu+\nu_m)}[\inf_{v^n\in V^h}\|\nabla(\bar{z}_{+,t}(t_{n+1})-v^n)\|^2+\inf_{v^n\in V^h}\|\nabla(\bar{z}_{-,t}(t_{n+1})-v^n)\|^2]
\end{aligned}
$$

$$
\begin{aligned}
&+\frac{\Delta t^2}{(\nu+\nu_m)}[\|\bar{z}_{+,tt}(t_{n+1})\|^2+\|\bar{z}_{-,tt}(t_{n+1})\|^2]\\
&+\frac{\Delta t^2\delta^4}{(\nu+\nu_m)}[\|\nabla\bar{z}_{+,tt}(t_{n+1})\|^2+\|\nabla\bar{z}_{-,tt}(t_{n+1})\|^2]\\
&+(\nu+\nu_m)\delta^4[\|\nabla\tilde{\zeta}_+(t_{n+1})\|^2+\|\nabla\tilde{\zeta}_-(t_{n+1})\|^2]\\
&+\frac{(\nu-\nu_m)^2}{(\nu+\nu_m)}\delta^4[\|\nabla\tilde{\zeta}_+(t_n)\|^2+\|\nabla\tilde{\zeta}_-(t_n)\|^2]\\
&+\frac{\Delta t^2}{(\nu+\nu_m)}[\|\nabla\bar{z}_+(t_{n+1})\|^2\|\nabla\bar{z}_{-,t}(t_{n+1})\|_\infty^2+\|\nabla\bar{z}_-(t_{n+1})\|^2\|\nabla\bar{z}_{+,t}(t_{n+1})\|_\infty^2]\\
&+[\inf_{q^n\in Q^h}\|p_+(t_{n+1})-q^n\|^2+\inf_{q^n\in Q^h}\|p_-(t_{n+1})-q^n\|^2]\Big)\Big\}
\end{aligned}
$$

Corollary 2. *If Taylor-Hood* (P_2, P_1) *pair of finite element spaces is used, the overall accuracy of the ADM is given by*

$$
\begin{aligned}
&\|(\bar{z}_+(t_N)-\omega_+^N)\|_{A_+}^2+\|(\bar{z}_-(t_N)-\omega_-^N)\|_{A_-}^2\\
&+2\Delta t\sum_{n=0}^{N-1}\Big\{\frac{\nu\nu_m}{(\nu+\nu_m)}[\|\nabla(\bar{z}_+(t_n)-\omega_+^n)\|^2+\|\nabla(\bar{z}_-(t_n)-\omega_-^n)\|^2]\\
&+\frac{(\nu+\nu_m)}{8}(\delta_+^2\|(\tilde{\zeta}_+(t_n)-\zeta_+^n)\|^2+\delta_-^2\|(\tilde{\zeta}_-(t_n)-\zeta_-^n)\|^2)]\Big\}\\
&\le C[\delta_+^4+\delta_-^4+\Delta t^2+h^4]
\end{aligned}
$$

The results of Theorems 23 and 24 are sufficient to prove the unconditional stability of the ADC solution $W_{+/-}$ – see [164].

Theorem 25. *Let* W_+, W_- *satisfy (10.4). Let* $f_+, f_- \in L^2(0,T;H^{-1}(\Omega))$. *Then there exists a constant* $C_1 = C_1(\nu,\nu_m)$, *s.t.*

$$
\begin{aligned}
&\|W_+^{n+1}\|^2+\delta_+^2\|\nabla W_+^{n+1}\|^2+\|W_-^{n+1}\|^2+\delta_-^2\|\nabla W_-^{n+1}\|^2 \qquad (10.6)\\
&+\min(\nu,\nu_m)\Delta t\sum_{i=1}^{n+1}(\|\nabla W_+^i\|^2+\|\nabla W_-^i\|^2)\\
&+2\min(\nu,\nu_m)\Delta t\sum_{i=1}^{n+1}(\delta_+^2\|Z_+^i\|^2+\delta_-^2\|Z_-^i\|^2)\\
&\le C_1(\|w_+^0\|^2+\delta_+^2\|\nabla w_+^0\|^2+\|w_-^0\|^2+\delta_-^2\|\nabla w_-^0\|^2\\
&+\Delta t\sum_{i=1}^{n+1}(\|f_+(t_i)\|_{-1}^2+\|f_-(t_i)\|_{-1}^2)).
\end{aligned}
$$

The authors of [164] stopped here, and didn't prove the accuracy of the correction step approximation. The proofs of Theorems 23, 24 and 25 are lengthy and tedious, but the interested reader could definitely take three extra steps, proving the accuracy of the gradient of the ADM error, the accuracy of the time derivative of the ADM error, and then (using these two results) prove the accuracy of the ADC solution.

10.4 COMPUTATIONAL TESTS

10.4.1 Problem with a Manufactured Solution

Two types of numerical tests were carried out in [164]. First, the authors set out to verify the claimed convergence rates. For a problem with a manufactured solution they chose $\delta = C_1\sqrt{h}$, $\Delta t = C_2 h$ (where h is the mesh diameter) and they did several mesh refinements. This is an honest way to test both the second order accuracy of the time discretization, and the fourth order of the modeling error: with this relationship between h, δ, and Δt we get $O(h^2 + (\Delta t)^2 + \delta^4) = O(h^2)$ - so the ADC rates should be approaching the second order. Indeed, in [164] this was shown to be the case for a flow with $\nu = 10^{-4}$, $\nu_m = 10^{-3}$.

10.4.2 Flow Past a Step: Magnetic Field is Induced by the Electrically Conducting Flow

Flow past a step is by now a familiar test problem; it was chosen in [164] with $Re = 600$, $Re_m = 700$, and with the time step $\Delta t = 0.05$. The mean magnetic field B_0 was taken to be zero, so that the magnetic fluctuations b would be induced by the electrically conducting flow itself. Note also that a large value of $Re_m = 700$ would force the velocity streamlines to be "frozen together" with the streamlines of the magnetic field, for any nonzero value of B_0. Thus, another viable setting for this test problem, that we encourage the reader to try, consists of a small but nonzero mean magnetic field (say, $B_0 =< 0.1, 0 >$) and a moderate value of the magnetic Reynolds number Re_m.

Figure 10.1 presents the flow at the final time $T = 40$, computed by the ADM (left) and the ADC (right) models. As was the case for the flow past a step in the NSE setting, we see vortices forming, shedding and traveling. The reference solution is given in Figure 10.2. It was computed by direct numerical simulation with the second order accurate Crank–Nicolson method for time discretization, at $\Delta t = 0.005$. We notice that, despite the added modeling error and despite the models' time step being ten times larger than the one used by the reference solution, the ADC was still able to correctly capture the qualitative features of the solution. The ADM failed to do so.

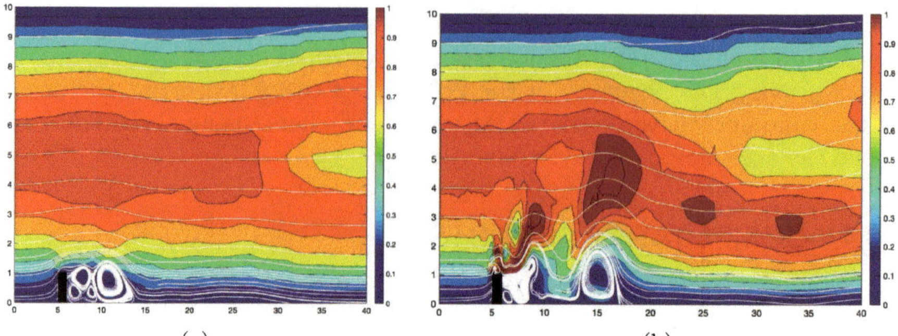

(a) (b)

Figure 10.1 Defect and Correction Step solutions at $T = 40$, $\Delta t = 0.05$.

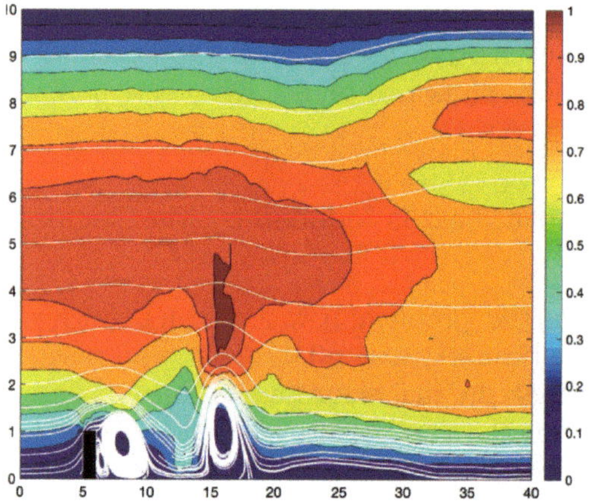

Figure 10.2 True solution, DNS velocity for MHD with $\Delta t = 0.005$, at $T = 40$.

CHAPTER 11

Another LES-C Model for a Fluid-Fluid Interaction Problem

11.1 FLUID–FLUID INTERACTION AT HIGH REYNOLDS NUMBERS: CHALLENGES IN CONSTRUCTING AN LES MODEL

In this chapter we will view another coupled Navier–Stokes system, with one or both flows at high Reynolds numbers. We aim to decouple the problem, find the right fit among the existing LES turbulence models, and then build and analyze the corresponding full LES-C model, that would reduce both the modeling error and the temporal discretization error of the LES, [177].

Let us return to the fluid–fluid interaction problem, discussed in Chapter 6. Consider the two-dimensional spatial domain Ω, comprised of two subdomains Ω_1 and Ω_2, coupled across an interface I, as shown in Figure 11.1.

The problem for the velocity-pressure pair (u, p) reads: *given* $\nu_i > 0$, $f_i : [0, T] \to H^1(\Omega_i)^d, u_i(0) \in H^1(\Omega_i)^d$ *and* $\kappa \in \mathbb{R}$, *find (for* $i = 1, 2$*)* $u_i : \Omega_i \times [0, T] \to \mathbb{R}^d$ *and* $p_i : \Omega_i \times [0, T] \to \mathbb{R}$ *satisfying (for* $0 < t \leq T$*)*

$$
\begin{aligned}
u_{i,t} = \nu_i \Delta u_i - u_i \cdot \nabla u_i - \nabla p_i + f_i, \quad & \text{in } \Omega_i, & (11.1)\\
-\nu_i \hat{n}_i \cdot \nabla u_i \cdot \tau = \kappa |u_i - u_j| (u_i - u_j) \cdot \tau \text{ on } I, \; & i, j = 1, 2, i \neq j & (11.2)\\
u_i \cdot \hat{n}_i = 0 \quad \text{on } I, \quad & i = 1, 2 & (11.3)\\
\nabla \cdot u_i = 0, \quad & \text{in } \Omega_i & (11.4)\\
u_i(x, 0) = u_i^0(x), \quad & \text{in } \Omega_i, & (11.5)\\
u_i = 0, \quad & \text{on } \Gamma_i = \partial\Omega_i \setminus I. & (11.6)
\end{aligned}
$$

The vectors $\hat{n}_i$ are the unit normals on $\partial\Omega_i$, and τ is any vector such that $\tau \cdot \hat{n}_i = 0$. The parameters ν_i represent kinematic viscosities. This problem

DOI: 10.1201/9781003355977-11

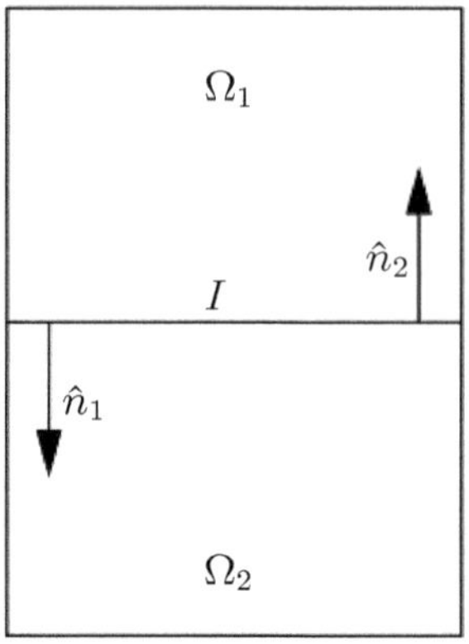

Figure 11.1 Example subdomains, coupled across an interface I.

could be viewed as the important first step in studying turbulent atmosphere–ocean problems. The nonlinear coupling equation (11.2), known as the rigid lid condition, was introduced in [130, 131] to allow for the exchange of energies between the subdomains, while having the global energy of the system conserved. This extra nonlinearity poses substantial difficulties in both the numerical analysis of the problem, and in formulating the turbulence models.

The need for decoupling was discussed in Chapter 6. The two main reasons – computational efficiency and the ability to use legacy codes for the subdomains – are still valid and even more pronounced in the case when one or both flows exhibit turbulent behavior. Atmosphere–ocean applications with turbulent flows include modeling of hurricanes and monsoons, weather prediction, and modeling of objects that fly over a water surface at high speed and low altitude.

11.1.1 Previous Results: The Interface Velocities Should not Be Filtered

The literature on turbulence modeling in fluid–fluid interaction problems is scarse; some results are mentioned in [177]. Some of the latest attempts to decouple the system first, and then apply an LES turbulence model were attempted in [180, 181]. In [180] the authors tried to build the ADM on top of the Geometric Averaging (GA) decoupling method of [118] (used also in Chapter 6). The somewhat surprising result was that the nonlinear coupling condition allowed for three different formulations of the ADM – all within the general framework of the approximate deconvolution models, but the results of these three models were quite different. The authors concluded in [180], that the best numerical results are obtained for the variation of the GA-ADM where the non-filtered velocity is used in the interface terms. This was also the only formulation, for which the authors could not prove stability. This highlighted the following issue: in order for a combination of an LES model and the GA method to be stable, the model has to maintain a certain structure of the nonlinear term.

A good fit was found in [181], where the authors combined the GA partitioning method with another LES model called NS-ω. They showed that this GA-NS-ω model is stable and reasonably accurate, as it uses the non-filtered velocity on the interface. In this chapter we will describe the full LES-C model, that is built on GA-NS-ω to reduce both the modeling error and the time discretization error. This NS-ω-C model (or GA-NS-ω-C, to account for the partitioning method) is investigated theoretically and numerically in [177].

The NS-ω model, proposed and studied in [182, 183], is given by

$$\begin{aligned} u_t + \bar{\omega} \times u - \nu \Delta u + \nabla q &= f, \\ \nabla \cdot u &= 0, \end{aligned} \tag{11.7}$$

where the filtered vorticity is defined by $\bar{\omega} = \nabla \times \bar{u}$, $\bar{u} - \delta^2 \Delta \bar{u} = u$.

Combining the Geometric Averaging method (see [118] or Chapter 6) with the NS-ω model leads to the GA-NS-ω formulation: find $(\hat{u}_i^{n+1}, \overline{\hat{u}}_i^{n+1}, \hat{q}_i^{n+1})$ in (X_i^h, X_i^h, Q_i^h), $i = 1, 2$, such that for any $(v_i, \xi_i, \chi_i) \in (X_i^h, X_i^h, Q_i^h)$

$$\begin{aligned} \left(\frac{\hat{u}_i^{n+1} - \hat{u}_i^n}{\Delta t}, v_i \right) &+ \nu_i \left(\nabla \hat{u}_i^{n+1}, \nabla v_i \right) \\ +\kappa \int_I |[\hat{u}^n]| \hat{u}_i^{n+1} v_i ds &- \kappa \int_I \hat{u}_j^n |[\hat{u}^n]|^{1/2} |[\hat{u}^{n-1}]|^{1/2} v_i ds \\ -(\hat{q}_i^{n+1}, \nabla \cdot v_i) + b_\omega \left(\overline{\hat{u}}_i^{n+1}, \hat{u}_i^{n+1}; v_i \right) &= \left(f_i(t^{n+1}), v_i \right), \ \forall v_i \in X_{i,h}, \\ \delta_i^2 (\nabla \overline{\hat{u}}_i^{n+1}, \nabla \xi_i) + (\overline{\hat{u}}_i^{n+1}, \xi_i) &= (\hat{u}_i^{n+1}, \xi_i), \ \forall \xi_i \in X_{i,h}, \\ (\nabla \cdot \hat{u}_i^{n+1}, \chi_i) &= 0, \ \forall \chi_i \in Q_{i,h}, \end{aligned} \tag{11.8}$$

where $b_\omega(u, v; w) = ((\nabla \times u) \times v, w)$, $\forall u, v, w \in X_{i,h}, i = 1, 2$.

11.2 THE NS-ω-BASED LES-C MODEL

In order to build the full LES-C model using (11.8) as the defect step, we identify the quantities of interest (NS-ω seeks the non-filtered velocity and pressure) and use Algorithm 7.1.1; then add deferred correction to the newly constructed model, to reduce the temporal discretization error of the correction step solution. The correction step of the full GA-NS-ω-C model [177], aiming to reduce the modeling error and the temporal discretization error of (11.8), seeks $(\tilde{u}_i^{n+1}, \overline{\tilde{u}}_i^{n+1}, \tilde{q}_i^{n+1})$ in (X_i^h, X_i^h, Q_i^h), $i = 1, 2$, such that for any $(v_i, \xi_i, \chi_i) \in (X_i^h, X_i^h, Q_i^h)$

$$\left(\frac{\tilde{u}_i^{n+1} - \tilde{u}_i^n}{\Delta t}, v_i \right) + \nu_i \left(\nabla \tilde{u}_i^{n+1}, \nabla v_i \right) \tag{11.9}$$

$$+b_\omega\left(\overline{\tilde{u}}_i^{n+1},\tilde{u}_i^{n+1};v_i\right)-(\tilde{q}_i^{n+1},\nabla\cdot v_i)$$
$$+\kappa\int_I|[\tilde{u}^n]|\tilde{u}_i^{n+1}v_i ds-\kappa\int_I\tilde{u}_j^n|[\tilde{u}^n]|^{1/2}|[\tilde{u}^{n-1}]|^{1/2}v_i ds$$
$$=\left(\frac{f_i(t^{n+1})+f_i(t^n)}{2},v_i\right)$$
$$+\frac{\nu_i\Delta t}{2}\left(\nabla\left(\frac{\hat{u}_i^{n+1}-\hat{u}_i^n}{\Delta t}\right),\nabla v_i\right)-\frac{\Delta t}{2}\left(\frac{\hat{q}_i^{n+1}-\hat{q}_i^n}{\Delta t},\nabla\cdot v_i\right)$$
$$+b_\omega\left(\overline{\hat{u}}_i^{n+1},\hat{u}_i^{n+1};v_i\right)-\frac{1}{2}b_\omega\left(\hat{u}_i^{n+1},\hat{u}_i^{n+1};v_i\right)-\frac{1}{2}b_\omega\left(\hat{u}_i^{n},\hat{u}_i^{n};v_i\right)$$
$$+\frac{\kappa}{2}\Delta t\int_I|[\hat{u}^n]|\left(\frac{\hat{u}_i^{n+1}-\hat{u}_i^n}{\Delta t}\right)v_i ds$$
$$-\frac{\kappa}{2}\Delta t\int_I\hat{u}_i^{n+1}\left(\frac{|[\hat{u}^{n+1}]|-|[\hat{u}^n]|}{\Delta t}\right)v_i ds$$
$$-\kappa\int_I\hat{u}_j^n|[\hat{u}^n]|^{1/2}|[\hat{u}^{n-1}]|^{1/2}v_i ds$$
$$+\frac{\kappa}{2}\int_I|[\hat{u}^{n+1}]|\hat{u}_j^{n+1}v_i ds+\frac{\kappa}{2}\int_I|[\hat{u}^n]|\hat{u}_j^n v_i ds,\ \forall v_i\in X_{i,h},$$
$$\delta_i^2(\nabla\overline{\tilde{u}}_i^{n+1},\nabla\xi_i)+(\overline{\tilde{u}}_i^{n+1},\xi_i)=(\tilde{u}_i^{n+1},\xi_i),\ \forall\xi_i\in X_{i,h},$$
$$(\nabla\cdot\tilde{u}_i^{n+1},\chi_i)=0,\ \forall\chi_i\in Q_{i,h},$$

The defect and correction steps of the GA-NS-ω-C model (11.8)–(11.9) enjoy the same structure of the left-hand side, and the model is amenable to parallelization: $\hat{u}_i^{n+1}$ and $\tilde{u}_i^n$ should be simultaneously computed on two cores.

11.3 STABILITY AND ACCURACY OF THE NEW MODEL

11.3.1 Defect Step Results

The stability and accuracy theorems, concerning the GA-NS-ω-C model, are derived in [177]. The authors had to overcome all the pitfalls typical of Chapters 7 (LES-C) and 6 (fluid–fluid interaction with nonlinear coupling). The notation $\|\cdot\|$ is used in the theorems below to denote norms with the following meaning

$$\|x\|=\begin{cases}\|x\|_{\Omega_1} & \text{if } x \text{ has support on } \Omega_1\\ \|x\|_{\Omega_2} & \text{if } x \text{ has support on } \Omega_2\end{cases}.$$

Stability and accuracy of the defect step solution (GA-NS-ω model) were established in [181].

Theorem 26 (Unconditional stability of the defect step). *Let* $(\hat{u}_i,\bar{\hat{u}}_i,q_i)\in(X_{i,h},X_{i,h},Q_{i,h})$ *be the solution of* (11.8), $i=1,2$. *Let also* $f_i\in L^2(0,T;H^{-1}(\Omega_i)),i=1,2$. *Then the solution is unconditionally stable and* *the following stability bound holds:*

$$\sum_{i=1}^{2}\left(\|\hat{u}_i^{n+1}\|^2+\nu_i\Delta t\sum_{k=1}^{n+1}\|\nabla\hat{u}_i^k\|^2\right)$$
$$+\frac{\kappa}{2}\int_I\left(|\hat{u}_1^{n+1}|^2+|\hat{u}_2^{n+1}|^2\right)|\hat{u}_1^n-\hat{u}_2^n|ds$$
$$+\frac{\kappa}{2}\Delta t\sum_{k=1}^{n}\int_I\left(|\hat{u}_1^k-\hat{u}_2^k|^{\frac{1}{2}}\hat{u}_1^{k+1}-\hat{u}_2^k|\hat{u}_1^{k-1}-\hat{u}_2^{k-1}|^{1/2}\right)^2ds$$
$$+\frac{\kappa}{2}\Delta t\sum_{k=1}^{n}\int_I\left(|\hat{u}_1^k-\hat{u}_2^k|^{\frac{1}{2}}\hat{u}_2^{k+1}-\hat{u}_1^k|\hat{u}_1^{k-1}-\hat{u}_2^{k-1}|^{1/2}\right)^2ds$$
$$\le\sum_{i=1}^{2}\left(\|\hat{u}_i^1\|^2+\frac{1}{4\nu_i}\Delta t\sum_{k=1}^{n+1}\|f_i(t^k)\|_{-1}^2\right)$$
$$+\frac{\kappa}{2}\int_I\left(|\hat{u}_1^1|^2+|\hat{u}_2^1|^2\right)|\hat{u}_1^0-\hat{u}_2^0|ds.$$

Under the time step restriction typical of the nonlinear NSE, the following accuracy result holds, [181].

Theorem 27 (Accuracy of the defect step solution). *Let* $(\hat{u}_i,\bar{\hat{u}}_i,q_i)\in(X_{i,h},X_{i,h},Q_{i,h})$ *be the solution of* (11.8), $i=1,2$. *Let also* $f_i\in L^2(0,T;H^{-1}(\Omega_i)),i=1,2$. *Assume also the following regularity of the true solution:* $u_i\in L^4(0,T;H^1(\Omega_i))$, $\bar{u}_i\in L^2(0,T;H^3(\Omega_i))$, $u_{i,t}\in L^2(0,T;H^1(\Omega_i))$, $u_{i,tt}\in L^2(0,T;L^2(\Omega_i))$. *Then, given the time step restriction* $\Delta t\le\frac{C\min\{\nu_1,\nu_2\}^3}{\kappa^4}$,

$$\sum_{i=1}^{2}\left(\|\hat{u}_i^{n+1}-u_i(t^{n+1})\|^2+\nu_i\Delta t\sum_{k=1}^{n}\|\nabla(\hat{u}_i^{k+1}-u_i(t^{k+1}))\|^2\right)$$
$$\le C\sum_{i=1}^{2}\left(\|\hat{u}_i^0-u_i(t^0)\|^2+\|\hat{u}_i^1-u_i(t^1)\|^2+\frac{\delta_i^4}{\nu_i}+\left(\frac{1}{\nu}+\frac{\kappa^2}{\nu^2}\right)\Delta t^2\right.$$
$$+\nu_i\|\nabla\left(\hat{u}_i^0-u_i(t^0)\right)\|^2+\nu_i\|\nabla\left(\hat{u}_i^1-u_i(t^1)\right)\|^2$$
$$+\frac{\Delta t}{\nu}\sum_{k=1}^{n}\left(\inf_{r_i^h\in X_i^h}\{\|\nabla\left(u_i(t^{k+1})-r_i^h\right)\|^2\}\right.$$
$$\left.\left.+\inf_{q_i^h\in Q_i^h}\{\|p_i(t^{k+1})-q_i^h\|^2\}\right)\right).$$

Next come the novel results of [177] - bounds on the derivatives of errors.

11.3.2 Accuracy of the Derivatives of the Defect Step Solution

Let $e_i^k=\hat{u}_i^k-u_i(t^k)$, $k=1,...,n$, $i=1,2$.

Theorem 28 (Accuracy of the time derivative of the defect step). *Let $(u_i, \bar{u}_i, q_i) \in (X_{i,h}, X_{i,h}, Q_{i,h})$ be the solution of (11.8), $i = 1,2$ and let the assumptions of Theorem 27 be satisfied. Let the true solution have the following regularity: $u(x,t) \in H^2(0,T;H^3(\Omega)) \cup H^3(0,T;L^2(\Omega))$. To treat the interface terms, additionally let $\min\{\Delta t, h\} \leq \nu\kappa^{-1}$, where $\nu = \min\{\nu_1, \nu_2\}$ and $\max\{\Delta t, h\} \leq \alpha C_f^{-1/4}$, where α is a constant such that $0 < \alpha < ||[u]||$ and C_f is the constant in Theorem 27. Then, $\Delta t < \kappa^2\nu^{-1}$ implies the following accuracy bound:*

$$\sum_{i=1}^{2}\left(\left\|\frac{e_i^{n+1}-e_i^n}{\Delta t}\right\|^2 + \nu_i \Delta t \sum_{k=1}^{n}\left\|\frac{\nabla(e_i^{k+1}-e_i^k)}{\Delta t}\right\|^2\right)$$
$$\leq C\Delta t \sum_{i=1}^{2}\sum_{k=1}^{n}\left(\delta^4 + \Delta t^2 + \left\|\frac{\nabla(e_i^1-e_i^0)}{\Delta t}\right\|^2\right.$$
$$\left. + \inf_{r\in X_i}\left\|\frac{u_i^{k+1}-u_i^k}{\Delta t} - r_i\right\|^2_{\Omega_i,-1} + \inf_{q\in Q_i}\left\|\frac{p_i^{k+1}-p_i^k}{\Delta t} - q_i\right\|^2\right)$$

The next result relates the filtered and non-filtered velocity errors. In a sense, it is the model's analogue of an inverse inequality. We will need to make some additional assumptions on the regularity of the true solution.

Assumption on the true solution: *There exist positive finite constants C_u, $C_{\nabla\times\overline{u}}$, $C_{[u]}$ and $C_{[u_{tt}]}$ where:*

$$C_u = \max_{i=1,2}\sup_{t\in[0,T]}\sup_{\mathbf{x}\in\Omega_i}|u_i(t,\mathbf{x})|$$
$$C_{\nabla\times\overline{u}} = \max_{i=1,2}\sup_{t\in[0,T]}\sup_{\mathbf{x}\in\Omega_i}|\nabla\times\overline{u}_i(t,\mathbf{x})|$$
$$C_{[u]} = \sup_{t\in[0,T]}\sup_{\mathbf{x}\in I}|[\mathbf{u}(t,\mathbf{x})]|$$
$$C_{[u_{tt}]} = \sup_{t\in[0,T]}\sup_{\mathbf{x}\in I}|[\mathbf{u}_{tt}(t,\mathbf{x})]|$$

Denote the error in the filtered velocity by
$g_i^{n+1} = \overline{u}_i^{n+1} - \overline{\hat{u}}_i^{n+1} = \overline{e}_i^{n+1}$.

Theorem 29 (Tighter accuracy bound on the filtered velocity for the defect step). *Let the conditions of Theorem 28 be satisfied. Additionally, for $i = 1,2$ assume $\delta_i^4 \leq \Delta t^2 < h$. Then, under the assumptions on the true solution, the following accuracy bound holds:*

$$\sum_{i=1}^{2}\left(\delta_i^{-2}\|e_i^{n+1}-g_i^{n+1}\|^2 + \|\nabla g_i^{n+1}\|^2\right.$$
$$\left. + \nu_i\delta_i^{-4}\sum_{k=2}^{n}\|e_i^{k+1}-g_i^{k+1}\|^2 + \nu_i\delta_i^{-2}\sum_{k=2}^{n}\|\nabla(e_i^{k+1}-g_i^{k+1})\|^2\right) \leq$$

$$\frac{C\kappa^2}{\nu^3}\sum_{i=1}^{2}\left(T\max_{l=0,\dots,n}\inf_{r_i\in X_i^h}\|\nabla(u_i^{l+1}-r_i)\|^2\right.$$
$$+T\max_{l=0,\dots,n}\inf_{q_i\in Q_i^h}\|p_i^{l+1}-q_i\|^2+\|e_i^0\|^2+\|e_i^1\|^2+\|\nabla e_i^0\|^2+\|\nabla e_i^1\|^2$$
$$\left.+\Delta t\sum_{k=2}^{n}\left(\delta_i^4+\Delta t^2+\inf_{r_i\in X_i^h}\|\nabla(u_i^{n+1}-r_i)\|^2+\inf_{v_i\in X_i^h}\|\nabla(\overline{u}_i^{n+1}-v_i)\|^2\right)\right)$$

11.3.3 Stability and Accuracy of the Correction Step

Finally, we present the results on stability and the increased accuracy of the correction step solution, [177].

Theorem 30 (Stability of the correction step approximation). *Let* $(\tilde{u},\overline{\tilde{u}},\tilde{q})\in(X_{i,h},X_{i,h},Q_{i,h})$. *Under the conditions of Theorem 29 the following bound holds:*

$$\sum_{i=1}^{2}\left(\|\hat{u}_i^{n+1}\|^2+\nu_i\Delta t\sum_{k=1}^{n+1}\|\nabla\hat{u}_i^k\|^2\right)$$
$$+\frac{\kappa}{2}\int_I\left(|\hat{u}_1^{n+1}|^2+|\hat{u}_2^{n+1}|^2\right)|\hat{u}_1^n-\hat{u}_2^n|ds$$
$$+\frac{\kappa}{2}\Delta t\sum_{k=1}^{n}\int_I\left(|\hat{u}_1^k-\hat{u}_2^k|^{\frac{1}{2}}\hat{u}_1^{k+1}-\hat{u}_2^k|\hat{u}_1^{k-1}-\hat{u}_2^{k-1}|^{1/2}\right)^2ds$$
$$+\frac{\kappa}{2}\Delta t\sum_{k=1}^{n}\int_I\left(|\hat{u}_1^k-\hat{u}_2^k|^{\frac{1}{2}}\hat{u}_2^{k+1}-\hat{u}_1^k|\hat{u}_1^{k-1}-\hat{u}_2^{k-1}|^{1/2}\right)^2ds$$
$$\le C\sum_{i=1}^{2}\left(\|\hat{u}_i^1\|^2+\frac{1}{4\nu_i}\Delta t\sum_{k=1}^{n+1}\left\|\frac{f_i(t^k+f_i(t^{k-1})}{2}\right\|_{-1}^2\right)$$
$$+\frac{\kappa}{2}\int_I\left(|\hat{u}_1^1|^2+|\hat{u}_2^1|^2\right)|\hat{u}_1^0-\hat{u}_2^0|ds.$$

The proof of the (increased) accuracy will be given in the next theorem. The spirit of it is explained by averaging the equations for the true solution across two adjacent time levels: for $i=1,2$ and for any $v_i\in X_i$ the true solution (u_i^{n+1}, p_i^{n+1}) satisfies

$$\begin{aligned}
&\left(\frac{u_i^{n+1}-u_i^n}{\Delta t}, v_i\right) + \nu_i(\nabla u_i^{n+1}, \nabla v_i) \\
&+ \kappa \int_I u_i^{n+1}|[u^n]|v_i\,ds - \kappa \int_I u_j^n|[u^n]|^{1/2}|[u^{n-1}]|^{1/2} v_i\,ds \\
&-(p_i^{n+1}, \nabla\cdot v_i) + b_\omega(\overline{u}_i^{n+1}, u_i^{n+1}; v_i) \\
&-\left(\frac{u_i^{n+1}-u_i^n}{\Delta t} - \frac{1}{2}\left(u_{i,t}^{n+1} + u_{i,t}^n\right), v_i\right) \\
&= \left(\frac{f_i(t^{n+1}) + f_i(t^n)}{2}, v_i\right) + \frac{1}{2}\left(\frac{\nabla(u_i^{n+1}-u_i^n)}{\Delta t}, v_i\right) \\
&+ b_\omega\left(\overline{u}_i^{n+1} - u_i^{n+1}, u_i^{n+1}, v_i\right) + \frac{1}{2} b_\omega\left(u_i^{n+1}, u_i^{n+1}; v_i\right) \\
&- \frac{1}{2} b_\omega\left(u_i^n, u_i^n; v_i\right) + \frac{\kappa}{2}\Delta t \int_I |[u^n]| \left(\frac{u_i^{n+1}-u_i^n}{\Delta t}\right) v_i\,ds \\
&- \frac{\kappa}{2}\Delta t \int_I u_i^{n+1}\left(\frac{|[u^{n+1}]| - |[u^n]|}{\Delta t}\right) v_i\,ds \\
&- \kappa \int_I u_j^n |[u^n]|^{1/2}|[u^{n-1}]|^{1/2} v_i\,ds \\
&+ \frac{\kappa}{2}\int_I u_j^{n+1}|[u^{n+1}]|v_i\,ds + \frac{\kappa}{2}\int_I u_j^n|[u^n]|v_i\,ds
\end{aligned} \tag{11.10}$$

Here one can notice the structure typical of a Crank–Nicolson scheme. Most of the terms have their counterparts in (11.9), which explains the reduction in the temporal discretization error. Similarly, the presence in (11.9) of the corresponding pairs to the nonlinear terms in the right-hand side of (11.10) is the reason for the reduced modeling error in GA-NS-ω-C.

Theorem 31 (Accuracy of the correction step approximation). *Let $(\tilde{u}, \overline{\tilde{u}}, \tilde{q}) \in (X_{i,h}, X_{i,h}, Q_{i,h})$. Let the conditions of Theorem 29 be satisfied. Then the following bound holds*

$$\begin{aligned}
&\sum_{i=1}^{2}\left(\|u_i^{n+1} - \tilde{u}_i^{n+1}\|^2 + \nu_i\|\nabla(u_i^{n+1} - \tilde{u}_i^{n+1})\|^2\right) \leq \\
&\qquad \frac{C\kappa^2}{\min\limits_{i=1,2}\nu_i^3}\sum_{i=1}^{2}\Bigg(T\delta_i^4 \max_{l=0,\dots,n}\inf_{r_i\in X_i^h}\|\nabla(u_i^{l+1} - r_i)\|^2 \\
&+ T\delta_i^4 \max_{l=0,\dots,n}\inf_{q_i\in Q_i^h}\|p_i^{l+1} - q_i\|^2 + \|\tilde{e}_i^0\|^2 + \|\tilde{e}_i^1\|^2 + \|\nabla\tilde{e}_i^0\|^2 + \|\nabla\tilde{e}_i^1\|^2 \\
&+ \Delta t\sum_{k=2}^{n}\left(\delta_i^8 + \delta_i^4\Delta t^2 + \Delta t^4 + \inf_{r_i\in X_i^h}\|\nabla(u_i^{n+1} - r_i)\|^2\right)\Bigg)
\end{aligned}$$

11.4 NOVEL TEST PROBLEMS FOR THE VALIDATION OF THE GA-NS-ω-C MODEL

We now discuss several numerical experiments, that the authors of [177] used to validate the model. The comparison will be made between the GA-NS-ω-C and its most natural competitor, the GA-NS-ω model. All computations in [177] were performed using the open source finite element library Deal.II, [185].

11.4.1 Convergence Rates

Let us view the convergence rates first; choose the problem data so that the true solution $u_i = (u_{1,i}, u_{2,i}), i = 1, 2$ in $\Omega = \Omega_1 \cup \Omega_2$ with $\Omega_1 = [0, 1] \times [0, 1]$ and $\Omega_2 = [0, 1] \times [0, -1]$ is given by

$$
\begin{aligned}
u_{1,1} &= a\nu_1 e^{-2t}(1-x)^2(1+y)x^2 + ae^{-t}(1-x)x\nu_1/\sqrt{\kappa a}\\
u_{1,2} &= a\nu_1 e^{-2t}(2+y)(1-x)(2x-1)xy + ae^{-t}(2x-1)y\nu_1/\sqrt{\kappa a}\\
u_{2,1} &= a\nu_1 e^{-2t}(1-x)^2(1+\frac{\nu_1}{\nu_2}y)x^2\\
u_{2,2} &= a\nu_1 e^{-2t}(1-x)(2x-1)(2+\frac{\nu_1}{\nu_2}y)xy.
\end{aligned}
$$

Take the coupling coefficient to be $\kappa = 1$, the final time $T = 1$, and test the models at different values of the viscosity coefficients. As the Reynolds numbers increase, we expect the convergence rates to decay, but the LES-C model should still outperform the LES model. In order to check the reduction of both the modeling and the time discretization errors, set $\Delta t = h = 1/N$, $\delta = \sqrt{h}/32$; as the N value gets reduced by a factor of two, we expect the linear convergence of the GA-NS-ω and the quadratic convergence of the GA-NS-ω-C.

Tables 11.1–11.3 paint an interesting picture. Indeed, for the most part the theoretical findings were verified: the LES-C model has higher convergence rates that the LES. This is, however, dependent upon the size of the viscosity coefficient: as ν is reduced, the overall error increases and the convergence rates eventually decay. This conclusion might seem discouraging (albeit expected); yet, we point out the importance of comparing the results within each table. Advocating for building LES-C models on top of the existing LES models, we notice that the overall errors on the finest mesh are smaller for the GA-NS-ω-C, than for the GA-NS-ω. This is true even in the case of very small viscosities, Table 11.3. In fact, if one is concerned mainly with the L^2-norms of the errors, then the LES-C allows for a coarser mesh (one less mesh refinement) than what is needed for the LES. Thus, even at the smallest tested values of the viscosities the LES-C allows to reach the desired accuracy at a fraction of the CPU time needed for the LES model. The superiority of the LES-C

TABLE 11.1 $\nu_1 = 0.5$, $\nu_2 = 0.1$, $a = 1$

	Defect Step, GA-NS-ω			
N	$\|u-\hat{u}\|_{L^2(0,T;L^2(\Omega))}$	Rate	$\|u-\hat{u}\|_{L^2(0,T;H^1(\Omega))}$	Rate
8	4.11953e-04	-	8.50472e-03	-
16	2.03524e-04	1.02	2.70832e-03	1.65
32	1.01785e-04	1.00	1.05270e-03	1.36
64	5.09369e-05	1.00	4.79119e-04	1.14
128	2.54945e-05	1.00	2.33153e-04	1.04
	Correction Step, GA-NS-ω-C			
N	$\|u-\tilde{u}\|_{L^2(0,T;L^2(\Omega))}$	Rate	$\|u-\tilde{u}\|_{L^2(0,T;H^1(\Omega))}$	Rate
8	1.01226e-04	-	7.67579e-03	-
16	1.51884e-05	2.73	1.97315e-03	1.96
32	2.96769e-06	2.36	5.01399e-04	1.98
64	7.01584e-07	2.08	1.26683e-04	1.98
128	1.77424e-07	1.98	3.19387e-05	1.99

model over its LES counterpart will be further highlighted in the qualitative tests below.

11.4.2 Flow Past a Circular Obstacle Close to the Interface

The authors of [177] introduced several test problems for (11.1)–(11.6). The first setting concerns a circular object flying at a high speed and low altitude over the water surface: how does the water domain affect the vortex street that should be developing in the wake? Does this vortex street create any

TABLE 11.2 $\nu_1 = 0.005$, $\nu_2 = 0.001$, $a = 1/\nu_1$

	Defect Step, GA-NS-ω			
N	$\|u-\hat{u}\|_{L^2(0,T;L^2(\Omega))}$	Rate	$\|u-\hat{u}\|_{L^2(0,T;H^1(\Omega))}$	Rate
8	1.94534e-03	-	5.21231e-02	-
16	8.10025e-04	1.26	1.72017e-02	1.60
32	3.96054e-04	1.03	5.51475e-03	1.64
64	1.97319e-04	1.01	2.46784e-03	1.16
128	9.84947e-05	1.00	1.22070e-03	1.02
	Correction Step, GA-NS-ω-C			
N	$\|u-\tilde{u}\|_{L^2(0,T;L^2(\Omega))}$	Rate	$\|u-\tilde{u}\|_{L^2(0,T;H^1(\Omega))}$	Rate
8	1.19799e-03	-	5.18072e-02	-
16	1.72664e-04	2.79	1.55092e-02	1.74
32	1.63580e-05	3.40	3.00804e-03	2.37
64	1.41591e-06	3.53	4.87807e-04	2.62
128	2.39417e-07	2.56	8.66997e-05	2.49

TABLE 11.3 $\nu_1 = 0.00005$, $\nu_2 = 0.00001$, $a = 1/\nu_1$

	Defect Step, GA-NS-ω			
N	$\|u-\hat{u}\|_{L^2(0,T;L^2(\Omega))}$	Rate	$\|u-\hat{u}\|_{L^2(0,T;H^1(\Omega))}$	Rate
8	2.28773e-03	-	6.93553e-02	-
16	9.67960e-04	1.24	3.77867e-02	0.88
32	4.59428e-04	1.08	2.02999e-02	0.90
64	2.25652e-04	1.03	1.09554e-02	0.89
128	1.11834e-04	1.01	6.19037e-03	0.82
	Correction Step, GA-NS-ω-C			
N	$\|u-\tilde{u}\|_{L^2(0,T;L^2(\Omega))}$	Rate	$\|u-\tilde{u}\|_{L^2(0,T;H^1(\Omega))}$	Rate
8	1.59673e-03	-	6.85926e-02	-
16	4.00720e-04	1.99	3.60697e-02	0.93
32	9.68945e-05	2.05	1.78679e-02	1.01
64	2.43767e-05	1.99	8.25683e-03	1.11
128	1.01257e-05	1.27	4.56130e-03	0.86

coherent structures in the lower domain? This is a novel test problem, so one must first create a reference solution. This solution should be free from the modeling errors of the GA-NS-ω(-C), it should have a small enough time step size, and, if possible, it should be free from the decoupling errors of the GA method.

The monolithic method was used in [177] to produce a reference solution: the coupled problem (11.1)–(11.6) was solved by direct numerical simulation (Galerkin finite element method) on a finer spatial mesh with smaller time step.

When available, the comparison between the following three approaches will be made: GA-NS-ω, GA-NS-ω-C, and the monolithic method MM-BDF2. The latter is given by: find $(u_{h,i}^{n+1}, p_{h,i}^{n+1}) \in (X_i^h, Q_i^h)$ satisfying

$$\begin{aligned}\left(\frac{3u_{h,i}^{n+1} - 4u_{h,i}^{n} + u_{h,i}^{n-1}}{2\Delta t}, v_{h,i}\right) &+ \nu_i(\nabla u_{h,i}^{n+1}, \nabla v_{h,i}) \qquad (11.11)\\ +b_i^*\left(u_{h,i}^{n+1}; u_{h,i}^{n+1}, v_i\right) &- (p_{h,i}^{n+1}, \nabla \cdot v_{h,i})\\ +(\nabla \cdot u_{h,i}^{n+1}, q_{h,i}) + \kappa\int_I |[\mathbf{u}_h^{n+1}]|[\mathbf{u}_h^{n+1}]v_{h,i}ds &= (f_i^{n+1}, v_{h,i}),\end{aligned}$$

for all $(v_{h,i}, q_{h,i}) \in (X_i^h, Q_i^h)$. The coupling number κ is proportional to the drag coefficient (see the derivation of the interface condition in [130]: equations (1.22)–(1.25) and, in particular, the formula for the horizontal wind force); thus, it would be appropriate to investigate the behaviour of the system (11.1)–(11.6) at different values of κ. In particular, large κ could be considered when modeling a high speed air flow over an ocean domain. Large values of κ might lead to a more intricate mutual behavior of the two fluid domains, as this strengthens the coupling between the ocean and the atmosphere; in particular, the monolithic method MM-BDF2 might fail for larger values of κ. We will investigate some of these scenarios below.

The flow in the top domain $\Omega_1 = 2.2 \times 4.1$, given by the inflow/outflow conditions

$$u_{1,1}(0, y, t) = u_{1,1}(2.2, y, t) = \frac{6}{0.41^2} \sin(\pi t/8) y(0.41 - y),$$
$$u_{1,2}(0, y, t) = u_{1,2}(2.2, y, t) = 0,$$

passes a circular obstacle centered at (0.2, 0.06) with radius 0.05. Homogeneous Dirichlet boundary conditions are enforced on the walls of the lower domain, except the interface, and on the top wall of the top domain. The fluid in both domains has zero initial velocities. The following values of the parameters were used in [177]: $\nu_1 = 1e - 03$, $\nu_2 = 1$ and final time $T = 8$. The value of κ is increased from Figure 11.2 to Figure 11.4. As the coupling between the domains strengthens, we see the MM-BDF2 fail at $\kappa > 1$.

Both the NS-ω and NS-ω-C models produce comparable results in both domains at $\kappa = 1$: they capture the change in the direction of the vortex street from horizontal to slant and the induced fluid motion in the lower domain, and this is validated by the MM-BDF2 method at $\kappa = 1$. However, as we increase the exchange of fluxes between the subdomains, we expect to see more structures appear and develop in the lower domain from Figure 11.2 to Figure 11.4. The LES-C model GA-NS-ω-C shows exactly this type of behavior, while the GA-NS-ω solution is much less convincing, creating very few fluid structures in the lower domain (Figures 11.3, 11.4).

11.4.3 Flow Past a Cliff

Another qualitative experiment, created in [177] compares the models' solutions and the reference solution of the MM-BDF2 method (11.11) for a flow over the backward-facing step (the step arises from the bottom of the lower domain to the mid-height in the upper domain). This could be a description of a coastal mountain, cliff, etc., in a real-life simulation; see [187, 180, 181] for similar problem setups.

Homogeneous Dirichlet boundary conditions are enforced on the step, the top wall of the top domain, and the vertical walls of the lower domain. Parabolic inflow $u_1(0, y, t) = (4(y - 1)(2 - y), 0)$ and outflow $u_1(8, y, t) = (y(2 - y)/2, 0)$ drive the flow in the top domain. The authors of [177] used $\nu_1 = 0.005$, $\nu_2 = 1$, $T = 15$, $\Delta t = 0.01$ and $h \approx 0.2$.

Figure 11.5 shows the speed contours, obtained by the GA-NS-ω, GA-NS-ω-C and MM-BDF2, at $\kappa = 1$. Their qualitative features seem in agreement with each other: a recirculation region appears behind the step in the upper domain and the exchange of fluxes leads to the creation of a vortex in the

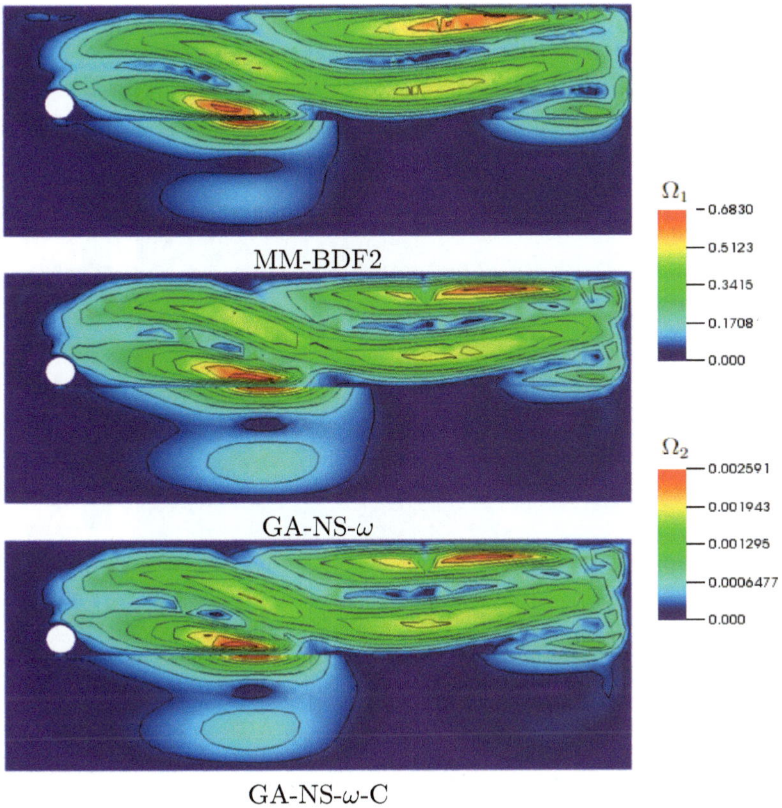

Figure 11.2 Expected velocity magnitudes with MM-BDF2, NS-ω and NS-ω-C for $\kappa = 1$.

lower domain underneath the recirculation region. This lower vortex spins in the opposite direction.

However, choosing $\kappa = 1000$ leads to significant differences in the models' behavior. Contrary to the flow past the obstacle at high κ, the MM-BDF2 method converges in this test problem and produces a reference solution at $\kappa = 1000$. The solution of the GA-NS-ω-C model is qualitatively similar to the reference solution (including the similar velocity magnitudes), while the solution of the GA-NS-ω comes with extra circulations in the lower domain, that are not observed in the monolithic solution (see Figure 11.6).

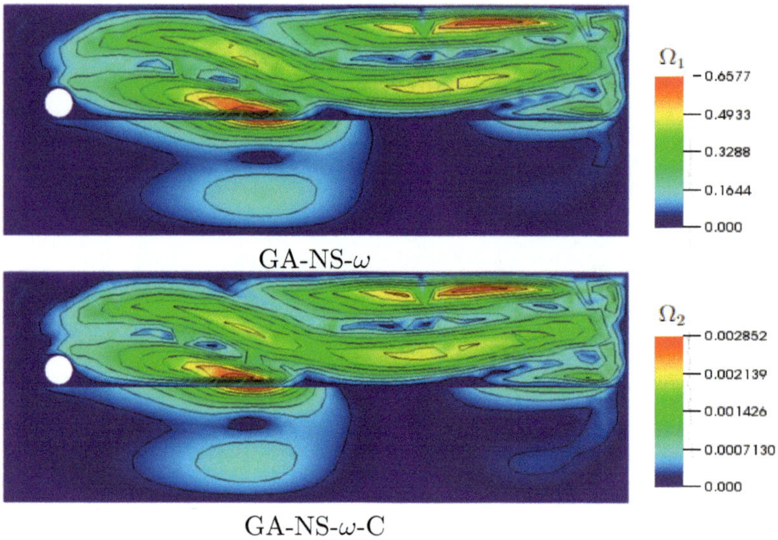

Figure 11.3 Expected velocity magnitudes with NS-ω and NS-ω-C for $\kappa = 10$. MM-BDF2 fails to converge.

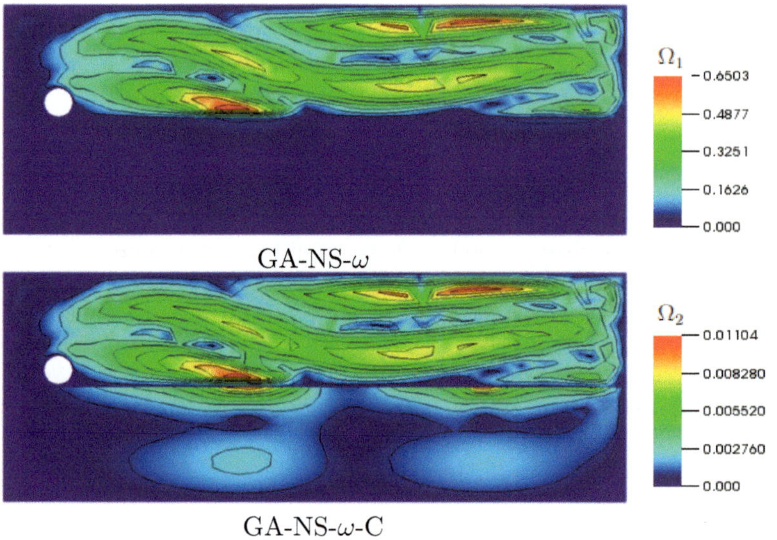

Figure 11.4 Expected velocity magnitudes with NS-ω and NS-ω-C for $\kappa = 1000$. MM-BDF2 fails to converge.

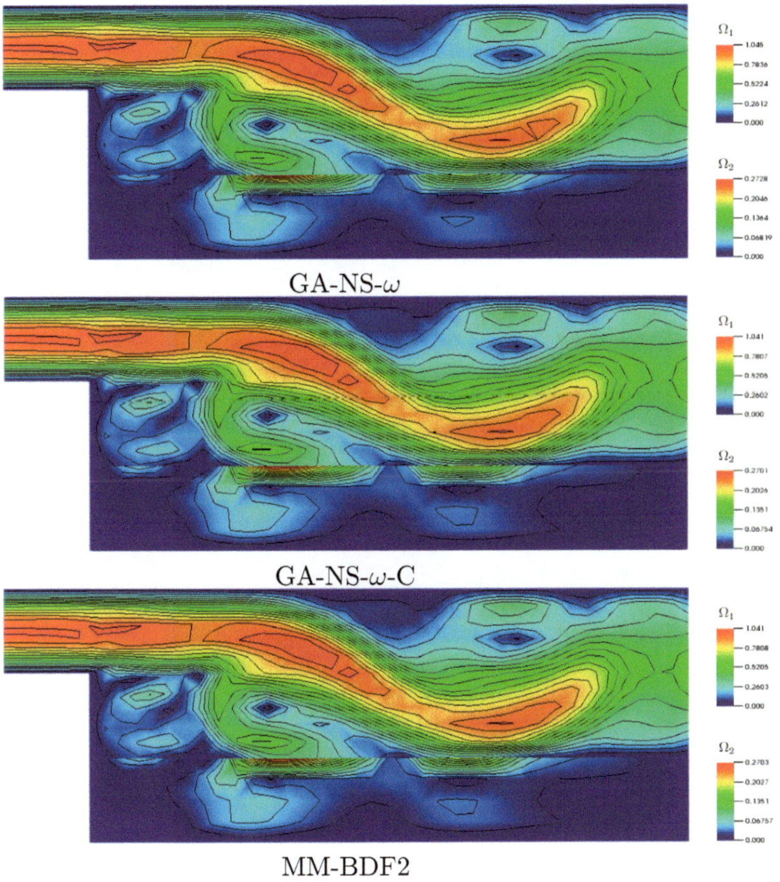

Figure 11.5 Expected velocity magnitudes for $\kappa = 1$.

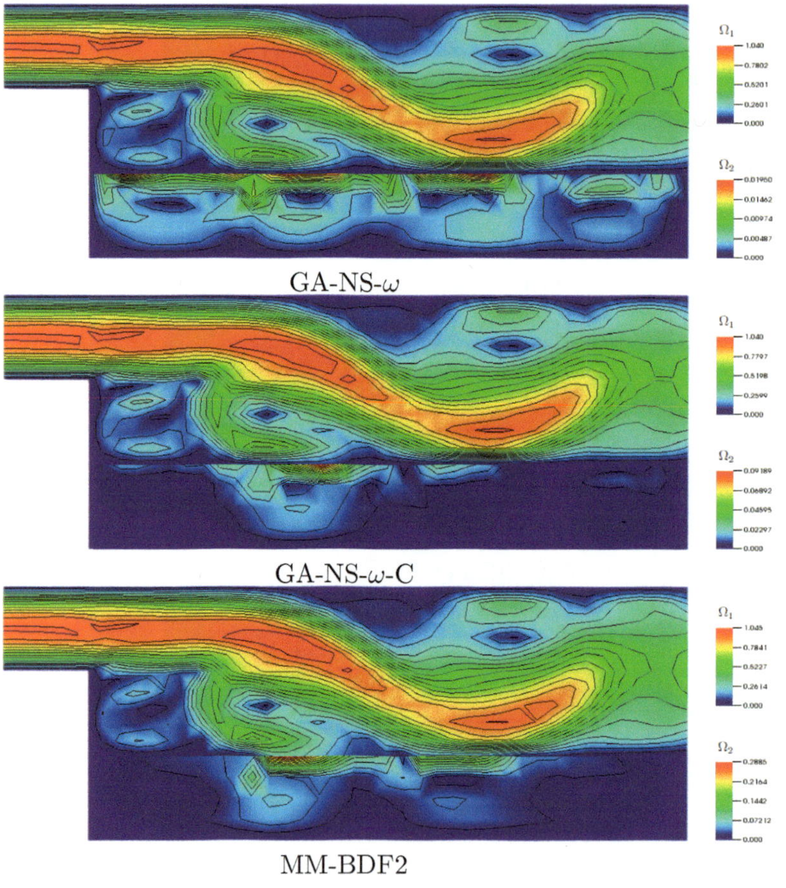

Figure 11.6 Expected velocity magnitudes for $\kappa = 1000$.

Bibliography

[1] G. I. Taylor, *The Spectrum of Turbulence*, Proceedings of the Royal Society A, vol. 164 (919), pp. 476–490, 1938.

[2] L. F. Richardson, *Weather Prediction by Numerical Processes*, Cambridge, University Press, 1922.

[3] Th. von Karman, *The Fundamentals of the Statistical Theory of Turbulence*, Journal of the Aeronautical Sciences, vol. 4(4), pp. 131–138, 1937.

[4] L. Fox, *Some Improvements in the Use of Relaxation Methods for the Solution of Ordinary and Partial Differential Equations*, Proceedings of the Royal Society of London. Series A, Mathematical and Physical Sciences Vol. 190, No. 1020, pp. 31–59, 1947.

[5] D.C. Wilcox, *Turbulence Modeling for CFD. 2nd Edition*, DCW Industries, La Canada, California, 1998.

[6] J. Leray, *Sur le mouvement d'un liquide visqueux emplissant l'espace*, Acta Mathematica, vol. 63, pp. 193–248, 1934.

[7] P. E. Zadunaisky, *A method for the estimation of errors propagated in the numerical solution of a system of ordinary differential equations*, The Theory of Orbits in the Solar System and in Stellar Systems. Proceedings from Symposium no. 25, 1964. Academic Press, London, p.281

[8] P.E. Zadunaisky, *On the Accuracy in the Numerical Computation of Orbits.* In: Giacaglia, G.E.O. (eds) Periodic Orbits, Stability and Resonances. Springer, Dordrecht, 1970.

[9] R. Frank, *The method of iterated defect-correction and its application to two-point boundary value problems*, Numer. Math. 25, pp. 409–419, 1975.

[10] R. Frank and C.W. Ueberhuber, *Iterated defect correction for the efficient solution of stiff systems of ordinary differential equations*, BIT 17, pp. 146–159, 1977.

[11] A. Labovsky, *A Defect Correction Method for the Time-Dependent Navier–Stokes Equations*, Numerical Methods for Partial Differential Equations, vol.25(1), pp. 1–25, 2008.

[12] K. Böhmer, P. W. Hemker, H. J. Stetter, The defect correction approach, in: K. Böhmer, H. J. Stetter (Eds.),*Defect Correction Methods. Theory and Applications*, Springer Verlag, 1984, pp. 1–32.

[13] P. W. Hemker, Mixed defect correction iteration for the accurate solution of the convection diffusion equation, pp. 485–501 in: *Multigrid Methods*, L.N.M. vol. 960, (W. Hackbusch and U. Trottenberg, eds.) Springer Verlag, Berlin 1982.

[14] P. W. Hemker, The use of defect correction for the solution of a singularly perturbed o.d.e., preprint, Mathematisch Centrum. Numerieke Wiskunde ; NW 139/82

[15] O. Axelsson and W. Layton, Defect correction methods for convection dominated, convection-diffusion equations, RAIRO J. Numer. Anal. 24 (1990) pp. 423-455.

[16] O. Axelsson and W. Layton, Optimal interior estimates for the defect-correction, finite element approach to 2-D convection-diffusion problems, ICMA report 88-116, Univ. of Pittsburgh, 1988.

[17] A. Bouaricha, Tensor-Krylov methods for Large Nonlinear Equations, Computational Optimization and Appl., 5 (1996), pp. 207-232.

[18] X.-C. Cai, Scalable nonlinear iterative methods for partial differential equations, LLNL report, Livermore, CA, 2000.

[19] W. Layton, H. K. Lee, J. Peterson, A defect-correction method for the incompressible Navier–Stokes equations, Applied Mathematics and Computation, Vol. 129, Issue 1, 2002, pp. 1-19.

[20] V. Ervin and W. Layton, High resolution minimal storage algorithms for convection dominated, convection diffusion equations, pp. 1173–1201 in: *Trans. of the Fourth Army Conf. on Appl. Math. and Comp.*, 1987

[21] V. Ervin and W. Layton, An analysis of a defect correction method for a model convection diffusion equation, SIAM J. Numer. Anal., 26 (1989), pp. 169–179.

[22] H. Elman, Y.-T. Shih, Iterative methods for stabilized discrete convection-diffusion problems, IMA J. Numer. Anal., 20 (2000), pp. 333-358.

[23] W. Heinrichs, Defect correction for convection dominated flow, SIAM J. Sci. Comput., 17 (1996), 1082-1091

[24] P. Hemker, An accurate method without directional bias for the numerical solution of a 2-D elliptic singular perturbation problem, pp. 192-206 in: *Theory And Applications Of Singular Perturbations*, Lecture Notes

in Math. 942, W. Eckhaus and E.M. de Jaeger, eds., Sprienger-Verlag, Berlin, 1982

[25] W. Layton, *Introduction to the Numerical Analysis of Incompressible Viscous Flows*, SIAM publications (Computational Science and Engineering Series), 2008.

[26] A. Labovschii *A defect correction method for the time-dependent Navier-Stokes equations*, Technical Report, University of Pittsburgh, 2007.

[27] J. Heywood, R. Rannacher, Finite-element approximations of the non-stationary Navier–Stokes problem. Part 4: Error analysis for second-order time discretization, SIAM J. Numer. Anal., Vol. 27, No. 2, pp. 353–384 (1990)

[28] B. Koren, Multigrid and Defect-Correction for the Steady Navier–Stokes Equations, *Applications to Aerodynamics*, C. W. I. Tract 74, Centrum voor Wiskunde en Informatica, Amsterdam, 1991.

[29] M.-H. Lallemand, B. Koren, Iterative defect correction and multigrid accelerated explicit time stepping schemes for the steady Euler equations, SIAM Journal on Scientific Computing, vol. 14(4), 1993, pp. 953–970.

[30] V. J. Ervin, H. K. Lee, Defect correction method for viscoelastic fluid flows at high Weissenberg number, Numerical Methods for Partial Differential Equations, Volume 22, Issue 1, pp. 145–164, 2006.

[31] P. W. Hemker, G. I. Shishkin, L. P. Shishkina, High-order time-accurate schemes for singularly perturbed parabolic convection-diffusion problems with Robin boundary conditions, Computational Methods in Applied Mathematics, Vol. 2 (2002), No. 1, pp. 3-25.

[32] J. H. Mathews, K. D. Fink, Numerical methods using MATLAB®. Pearson Prentice Hall, 2004.

[33] M. L. Minion, Semi-Implicit Projection Methods for Incompressible Flow based on Spectral Deferred Corrections, Appl. Numer. Math., 48(3-4), 369-387, 2004

[34] A. Bourlioux, A. T. Layton, M. L. Minion, High-Order Multi-Implicit Spectral Deferred Correction Methods for Problems of Reactive Flows, Journal of Computational Physics, Vol. 189, No. 2, pp. 651-675, 2003.

[35] W. Kress, B. Gustafsson, Deferred Correction Methods for Initial Boundary Value Problems, Journal of Scientific Computing, Springer Netherlands, Vol. 17, No. 1-4, 2002.

[36] A. Dutt, L. Greengard, V. Rokhlin, Spectral deferred correction methods for ordinary differential equations, BIT 40 (2), pp. 241-266, 2000.

[37] Y. Saad, Iterative Methods for Sparse Linear Systems. Second edition, published by SIAM, 2003.

[38] H. Alfvén, *Existence of electromagnetic-hydrodynamic waves*, Nature, 150 (1942), p. 405.

[39] L. Barleon, V. Casal, and L. Lenhart, *MHD flow in liquid-metal-cooled blankets*, Fusion Engineering and Design, 14 (1991), pp. 401–412.

[40] J. D. Barrow, R. Maartens, and C. G. Tsagas, *Cosmology with inhomogeneous magnetic fields*, Phys. Rep., 449 (2007), pp. 131–171.

[41] D. Biskamp, *Nonlinear magnetohydrodynamics*, vol. 1 of Cambridge Monographs on Plasma Physics, Cambridge University Press, Cambridge, 1993.

[42] D. Biskamp, *Magnetohydrodynamic turbulence*, Cambridge: Cambridge University Press, 2003.

[43] P. Bodenheimer, G. P. Laughlin, M. Różyczka, and H. W. Yorke, *Numerical methods in astrophysics*, Series in Astronomy and Astrophysics, Taylor & Francis, New York, 2007.

[44] P. A. Davidson, *An introduction to magnetohydrodynamics*, Cambridge Texts in Applied Mathematics, Cambridge University Press, Cambridge, 2001.

[45] M. Dobrowolny, A. Mangeney, and P. Veltri, *Fully developed anisotropic hydromagnetic turbulence in interplanetary space*, Phys. Rev. Lett., 45 (1980), pp. 144–147.

[46] C. R. Doering and J. D. Gibbon, *Applied analysis of the Navier–Stokes equations*, Cambridge Texts in Applied Mathematics, Cambridge University Press, Cambridge, 1995.

[47] E. Dormy and M. Núñez, *Introduction [Special issue: Magnetohydrodynamics in astrophysics and geophysics]*, Geophys. Astrophys. Fluid Dyn., 101 (2007), p. 169.

[48] E. Dormy and A. M. Soward, eds., *Mathematical aspects of natural dynamos*, vol. 13 of Fluid Mechanics of Astrophysics and Geophysics, Grenoble Sciences. Universite Joseph Fourier, Grenoble, 2007.

[49] W. Elsässer, *The hydromagnetic equations*, Phys. Rev., II. Ser., 79 (1950), p. 183.

[50] A. Fierros Palacios, *The Hamilton-type principle in fluid dynamics*, Springer, Vienna, 2006. Fundamentals and applications to magnetohydrodynamics, thermodynamics, and astrophysics.

[51] J. A. Font, *General relativistic hydrodynamics and magnetohydrodynamics: hyperbolic systems in relativistic astrophysics*, in Hyperbolic problems: theory, numerics, applications, Springer, Berlin, 2008, pp. 3–17.

[52] S. Galtier, S. V. Nazarenko, A. C. Newell, and A. Pouquet, *A weak turbulence theory for incompressible MHD*, 2000.

[53] J.-F. Gerbeau, C. Le Bris, and T. Lelièvre, *Mathematical methods for the magnetohydrodynamics of liquid metals*, Numerical Mathematics and Scientific Computation, Oxford University Press, Oxford, 2006.

[54] J. F. Gibson, J. Halcrow, and P. Cvitanović, *Equilibrium and travelling-wave solutions of plane Couette flow*, Journal of Fluid Mechanics, 638 (2009), pp. 243–266.

[55] P. Goldreich and S. Sridhar, *Toward a theory of interstellar turbulence. 2: Strong alfvenic turbulence*, Astrophysical Journal, 438 (1995), pp. 763–775.

[56] M. D. Gunzburger, O. A. Ladyzhenskaya, and J. S. Peterson, *On the Global Unique Solvability of Initial-Boundary Value Problems for the Coupled Modified Navier–Stokes and Maxwell Equations*. J. Math. Fluid Mech. 6, pp. 462–482, 2004.

[57] M. Gunzburger and A. Labovsky, *High Accuracy Method for Turbulent Flow Problems*, M3AS: Mathematical Models and Methods in Applied Sciences, vol. 22 (6), 2012, DOI: 10.1142/S0218202512500054.

[58] Hidetoshi and Hashizume, *Numerical and experimental research to solve MHD problem in liquid blanket system*, Fusion Engineering and Design, 81 (2006), pp. 1431 – 1438.

[59] W. Hillebrandt and F. Kupka, eds., *Interdisciplinary aspects of turbulence*, vol. 756 of Lecture Notes in Physics, Springer-Verlag, Berlin, 2009.

[60] P. S. Iroshnikov, *Turbulence of a conducting fluid in a strong magnetic field*, Soviet Astronom. AJ, 7 (1964), pp. 566–571.

[61] S. Y. Kadioglu, R. Klein, and M. L. Minion, *A fourth-order auxiliary variable projection method for zero-Mach number gas dynamics*, J. Comput. Phys., 227 (2008), pp. 2012–2043.

[62] R. Kraichnan, *Inertial-range spectrum of hydromagnetic turbulence*, Physics of Fluids, 8 (1965), p. 1385.

[63] L. D. Landau and E. M. Lifshitz, *Electrodynamics of continuous media*, Course of Theoretical Physics, Vol. 8. Translated from the Russian by J. B. Sykes and J. S. Bell, Pergamon Press, Oxford, 1960.

[64] W. Layton, H. Tran, and C. Trenchea, *Stability of partitioned methods for magnetohydrodynamics flows at small magnetic Reynolds number*, in Recent Advances in Scientific Computing and Applications: Proceedings of the 8th International Conference on Scientific Computing and Applications, E. M. Jichun Li and H. Yang, eds., vol. 586, AMS Contemporary Mathematics, 2013.

[65] W. Layton and C. Trenchea, *Stability of two IMEX methods, CNLF and BDF2-AB2, for uncoupling systems of evolution equations*, Applied Numerical Mathematics, 62 (2012), pp. 112–120.

[66] T. Lin, J. Gilbert, R. Kossowsky, and P. S. U. S. COLLEGE., *Sea-Water Magnetohydrodynamic Propulsion for Next-Generation Undersea Vehicles*, Defense Technical Information Center, 1990.

[67] E. Marsch, *Turbulence in the solar wind*, in Reviews in Modern Astronomy, G. Klare, ed., Springer, Berlin, 1990, p. 43.

[68] B. Punsly, *Black hole gravitohydromagnetics*, vol. 355 of Astrophysics and Space Science Library, Springer-Verlag, Berlin, second ed., 2008.

[69] P. H. Roberts, *An introduction to magnetohydrodynamics*, Elsevier, 1967.

[70] J. V. Shebalin, W. H. Matthaeus, and D. Montgomery, *Anisotropy in MHD turbulence due to a mean magnetic field*, Journal of Plasma Physics, 29 (1983), pp. 525–547.

[71] R. Temam, *Navier–Stokes Equations. Theory and Numerical Analysis*, North-Holland, Amsterdam, 1979.

[72] R. Temam, *Navier–Stokes Equations and Nonlinear Functional Analysis*, CBMS-NSF Regional Conference Series in Applied Mathematics, Society for Industrial and Applied Mathematics, Philadelphia, 1995.

[73] C. Trenchea, *Unconditional stability of a partitioned IMEX method for magnetohydrodynamic flows*, Applied Mathematics Letters, Vol. 27, 2014, pp. 97–100.

[74] M. K. Verma, *Statistical theory of magnetohydrodynamic turbulence: Recent results*, Physics Reports, 401 (2004), p. 229.

[75] I. Veselovsky, *Turbulence and waves in the solar wind formation region and the heliosphere*, Astrophys. Space Sci., 277 (2001), pp. 219–224.

[76] A. Labovsky, C. Trenchea, N. Wilson, *High Accuracy Method for Magnetohydrodynamics System in Elsasser Variables*, Computational Methods in Applied Mathematics, vol 15(1), pp. 97-110, 2015.

[77] W. Layton and M. Neda, *A similarity theory of approximate deconvolution models of turbulence*, Journal of Mathematical Analysis and Applications, vol. 333(1), 2007, pp. 416–429.

[78] W. Layton and L. Rebholz, *Approximate Deconvolution Models of Turbulence. Analysis, Phenomenology and Numerical Analysis*, Lecture Notes in Mathematics, vol. 2042, Springer Berlin, Heidelberg, 2012.

[79] M. Germano, *Differential filters for the large eddy numerical simulation of turbulent flows* , Phys. Fluids 29(6), 1986, pp. 1755–1757.

[80] N.A. Adams and S. Stolz, Deconvolution methods for subgrid-scale approximation in large-eddy simulation, in: *Modern Simulation Strategies for Turbulent Flow*, 2001, B. Geurts (editor), pp. 21-41, R. T. Edwards.

[81] S. Stolz, N. A. Adams and L. Kleiser, The Approximate Deconvolution Model for Compressible Flows: Isotropic Turbulence and Shock-Boundary-Layer Interaction , in: *Fluid Mechanics and Its Applications. Advances in LES of Complex Flows*, vol. 65, 2006, pp. 33-47, Springer Netherlands.

[82] C. Cardoso Manica and S. K. Merdan, Finite element error analysis of a zeroth order approximate deconvolution model based on a mixed formulation, JMAA, Vol. 331, No. 1, pp. 669–685, 2007.

[83] W. Layton and R. Lewandowski, Residual stress of approximate deconvolution models of turbulence, Journal of Turbulence, Vol. 7, Issue 46, p. 1–21, 2006.

[84] A. Labovsky, C. Trenchea, *Large Eddy Simulation for Turbulent Magnetohydrodynamic Flows*, Journal of Mathematical Analysis and Applications, 377 (2011), pp. 516–533.

[85] A. Labovsky, C. Trenchea, *A family of Approximate Deconvolution Models for MagnetoHydroDynamic Turbulence*, Numerical Functional Analysis and Optimization, vol.31(12), pp. 1362–1385, 2010.

[86] A. Labovsky, W.J. Layton, C.C. Manica, M. Neda, L.G. Rebholz, *The stabilized extrapolated trapezoidal finite-element method for the Navier–Stokes equations*, Computer Methods in Applied Mechanics and Engineering, vol. 198, Issues 9-12, 2009, pp. 958–974.

[87] A. Dunca and Y. Epshteyn, On the Stolz-Adams deconvolution model for the large eddy simulation of turbulent flows, SIAM J. Math. Anal., Vol. 37(6), pp. 1890–1902, 2006.

[88] M. Gunzburger, *Finite element methods for viscous incompressible flows: A guide to theory, practice, and algorithms*, Academic Press, Boston, 1989.

[89] W. Layton, L. Rebholz, C. Trenchea, Modular Nonlinear Filter Stabilization of Methods for Higher Reynolds Numbers Flow, University of Pittsburgh, technical report, 2010.

[90] V. Girault, P.A. Raviart, Finite element approximation of the Navier–Stokes equations, *Lecture notes in mathematics*, no. 749, Springer-Verlag, 1979.

[91] M. L. Minion, Semi-Implicit Projection Methods for Ordinary Differential Equations, Comm. Math. Sci., 1(3), 471–500, 2003.

[92] A.N. Kolmogorov, *The local structure of turbulence in incompressible viscous fluid for very large Reynolds number*, Dokl. Akad. Nauk. SSSR, vol. 30, 1941, pp. 9–13.

[93] Sagaut, P., Large eddy simulation for incompressible flows,*Scientific Computation*, Springer Verlag, 2006.

[94] John, V., Large eddy simulation of turbulent incompressible flows,*Lecture Notes in Computational Science and Engineering*, Springer Verlag, 2004.

[95] Berselli, L.C. and Iliescu, T. and Layton, W.J., Mathematics of large eddy simulation of turbulent flows,*Scientific Computation*, Springer Verlag, 2006.

[96] U. Piomelli and E. Balaras, *Wall-layer models for large eddy simulations*, Annu. Rev. Fluid Mech., vol. 34, 2002, pp. 349–372.

[97] J. Borggaard and T. Iliescu, Approximate deconvolution boundary conditions for large eddy simulation, *Applied Mathematics Letters*, vol. 19(8), 2006, pp. 735-740.

[98] P. Moin and M. Wang, Wall modeling for large-eddy simulation of turbulent boundary layers, IUTAM Symposium on One Hundred Years of Boundary Layer Research, *Solid Mechanics and Its Applications*, 2006, Volume 129, Session 5, 269-278.

[99] S.Y. Kadioglu, R. Klein and M.L. Minion, A fourth-order auxiliary variable projection method for zero-Mach number gas dynamics, *J. Comp. Physics*, 2008, Volume 227, 2012-2043.

[100] A. Labovsky, *A defect correction approach to turbulence modeling*, Numerical Methods for Partial Differential Equations, vol 31(1), pp. 268–288, 2015.

[101] H. J. Stetter, *The defect correction principle and discretization methods*, Numerische Mathematik, vol. 29(4), pp. 425-443, 1978.

[102] V. Ervin, W. Layton, J. Maubach, Adaptive defect correction methods for viscous incompressible flow problems, SIAM J. Numer. Anal., 37 (2000), pp. 1165-1185.

[103] W. Layton, L. Tobiska, *A Two-Level Method with Backtracking for the Navier–Stokes Equations*, SIAM JNA, 35, (1998), 2035-2054.

[104] S. C. Brenner and L. R. Scott, *The Mathematical Theory of Finite Element Methods*, Springer, New York, 2008.

[105] M. Aggul, J. Connors, D. Erkmen, and A. Labovsky, *A Defect-Deferred Correction Method for Fluid-Fluid Interaction*, SIAM Journal on Numerical Analysis, Vol. 56(4), 2018, pp. 2484–2512.

[106] R.A. Adams and J.J.F. Fournier, *Sobolev Spaces*, vol. 140 of Pure and Applied Mathematics, Academic Press, 2003.

[107] U. M. Ascher, S. J. Ruuth and B. T. R. Wetton, Implicit-explicit methods for time-dependent partial differential equations, SIAM J. Num. Anal. 32(3), 1995.

[108] G. Bellon, A. H. Sobel and J. Vialard, *Ocean-atmospere coupling in the monsoon intraseasonal oscillation: A simple model study*, Journal of Climate, Vol. 21, 2008, pp. 5254-5270.

[109] C. Bernardi, T. Chacon-Rebello M. Gomez, R. Lewandowski, F. Murat, A model of two coupled turbulent fluids, Part II: Numerical approximations by spectral discretization, SIAM Jour. Num. Analysis, Vol. 40, No. 6, pp. 2368-2394, 2002.

[110] H. Blum, S. Lisky and R. Rannacher, A Domain Splitting Algorithm for Parabolic Problems, Computing 49, 11–23, 1992.

[111] D. Bresch and J. Koko, Operator-Splitting and Lagrange Multiplier Domain Decomposition Methods for Numerical Simulation of Two Coupled Navier–Stokes Fluids, Int. J. Appl. Math. Comput. Sci., Vol. 16, No. 4, 2006, pp. 419–429.

[112] C. L. Brossier, V. Ducrocq and H. Giordani, *Effects of the air-sea coupling time frequency on the ocean response during Mediterranean intense events*, Ocean Dynamics, Vol. 59 (2009), pp. 539–549.

[113] E. Burman, Miguel A. Fernández, Stabilization of explicit coupling in fluid-structure interaction involving fluid incompressibility, Comput. Methods Appl. Mech. Engrg. 198 (2009), pp. 766–784.

[114] E. Burman, P. Hansbo, Interior penalty stabilized Lagrange multiplier for the finite element solution of elliptic interface problems, to appear in IMA Journal of Numerical Analysis, available online April 7, 2009.

[115] R. B. Neale, *et. al.*, *Description of the NCAR community atmosphere model (CAM 5.0)*, NCAR Tech. Note NCAR/TN-486+ STR (2010).

[116] P. Causin, J.-F. Gerbeau and F. Nobile, Added-mass effect in the design of partitioned algorithms for fluid-structure problems, Comput. Meth. Appl. Mech. Engrg., Vol. 194, No. 42-44, 2005, pp. 4506-4527.

[117] J. Connors, J. Howell and W. Layton, *Partitioned time stepping for a parabolic two domain problem*, SIAM Jour. Num. Anal., Vol. 47, No. 5, 2009, pp. 3526–3549.

[118] J. Connors, J. Howell and W. Layton, *Decoupled time stepping methods for fluid-fluid interaction*, SIAM Jour. Num. Analysis, Vol. 50, No. 3, 2012, pp. 1297-1319.

[119] J. Connors and B. Ganis, *Stability of algorithms for a two domain natural convection problem and observed model uncertainty*, Computational Geosciences, Vol. 15, No. 3 (2011), pp. 509-527.

[120] C. N. Dawson and Q. Du, A Finite Element Domain Decomposition Method for Parabolic Equations, Fourth International Symposium on Domain Decomposition Methods for PDEs, edited by R. Glowinski et al, pp. 255-263, SIAM, Philadelphia, 1991.

[121] S. D. Nicholls and S. G. Decker, *Impact of coupling an ocean model to WRF nor'easter simulations*, Monthly Weather Review 143.12 (2015), pp. 4997-5016.

[122] F. Lemarie and E. Blayo and L. Debreu, Analysis of ocean-atmosphere coupling algorithms: Consistency and stability, Procedia Comput. Sci., 51, pp. 2066-2075, 2015.

[123] F. Lemarié, P. Marchesiello, L. Debreu and E. Blayo, *Sensitivity of ocean-atmosphere coupled models to the coupling method: example of tropical cyclone Erica*, Doctoral dissertation, INRIA Grenoble; INRIA, 2014.

[124] F. Hecht, A. LeHyaric and O. Pironneau. Freefem++ version 2.24-1, 2008. `http://www.freefem.org/ff++`.

[125] Randall, D.A., R.A. Wood, S. Bony, R. Colman, T. Fichefet, J. Fyfe, V. Kattsov, A. Pitman, J. Shukla, J. Srinivasan, R.J. Stouffer, A. Sumi and K.E. Taylor, 2007: Climate Models and Their Evaluation. In: Climate Change 2007: The Physical Science Basis. Contribution of Working Group I to the Fourth Assessment Report of the Intergovernmental Panel on Climate Change [Solomon, S., D. Qin, M. Manning, Z. Chen, M. Marquis, K.B. Averyt, M.Tignor and H.L. Miller (eds.)]. Cambridge University Press, Cambridge, United Kingdom and New York, NY, USA.

[126] M. Jochum, G. Danabasoglu, M. Holland, Y.-O. Kwon and W. G. Large, *Ocean viscosity and climate*, Journal of Geophysical Research, Vol. 113, C06017, 2008, pp. 1-24.

[127] A. Labovsky, *A Defect Correction Method for the Evolutionary Convection Diffusion Problem with Increased Time Accuracy*, Computational Methods in Applied Mathematics, vol.9(2), pp. 154–164, 2009.

[128] Large, W.G., McWilliams, J.C. and Doney, S.C., *Oceanic vertical mixing: A review and a model with a nonlocal boundary layer*, Reviews of Geophysics, Vol. 32 (1994), pp. 363-403.

[129] R. Lewandowski, *Analyse Mathématique et Océanographic*, collection RMA, Masson, 1997.

[130] J.-L. Lions, R. Temam and S. Wang, *Models of the coupled atmosphere and ocean (CAO I)*, Computational Mechanics Advances, Vol. 1 (1993), pp. 5-54.

[131] J.-L. Lions, R. Temam and S. Wang, *Numerical analysis of the coupled atmosphere ocean models (CAO II)*, Computational Mechanics Advances, Vol. 1 (1993), pp. 55-119.

[132] M. L. Minion, *Semi-Implicit Spectral Deferred Correction Methods for Ordinary Differential Equations*, Comm. Math. Sci. 1 (2003), 471–500.

[133] P. A. Mooney, D. O. Gill, F. J. Mulligan and C. L. Bruyère, *Hurrican simulation using different representations of atmosphere-ocean interaction: the case of Irene*, Atmospheric Science Letters, Vol. 17, No. 7, 2016, pp. 415-421.

[134] J. Nelson, R. He, J. C. Warner and J. Bane, *Air-sea interactions during strong winter extratropical storms*, Ocean Dynamics, Vol. 64, No. 9, 2014, pp. 1233-1246.

[135] N. Perlin and E. D. Skyllingstad and R. M. Samelson and P. L. Barbour, Numerical simulation of air-sea coupling during coastal upwelling, J. Phys. Oceanography, 37, pp. 2081-2093, 2007.

[136] R. Smith, et. al., *The parallel ocean program (POP) reference manual*, Tech. Rep. LAUR-01853 141 (2010): 1-140.

[137] M. Aggul, A. Labovsky, *A High Accuracy Minimally Invasive Regularization Technique for Navier–Stokes Equations at High Reynolds Number*, Numerical Methods for Partial Differential Equations, 33(3), pp. 814-839, 2017.

[138] D. Erkmen, A. Labovsky, *Defect-deferred correction method for the two-domain convection-dominated convection-diffusion problem*, Journal of Mathematical Analysis and Applications, 450 (2017), pp. 180–196.

[139] B. Wang and X. Xie, *Coupled modes of the warm pool climate system. Part 1: The role of air-sea interaction in maintaining Madden-Julian Oscillation*, Journal of Climate, Vol. 11, 1998, pp. 2116-2135.

[140] W. L. Washington and C. L. Parkinson, *An Introduction to Three-Dimensional Climate Modeling: Second Edition*, University Science Books, Sausalito, California, 2005.

[141] R. M. Yablonsky, I. Ginis., *Limitation of one-dimensional ocean models for coupled hurricane–ocean model forecasts*, Monthly Weather Review 137.12 (2009): 4410-4419.

[142] Y. Zhang, Y. Hou and L. Shan, *Stability and convergence analysis of a decoupled algorithm for a fluid-fluid interaction problem*, SIAM Journal on Numerical Analysis, Vol. 54, No. 5 (2016), pp. 2833-2867.

[143] A. Labovsky, *Approximate Deconvolution with Correction – a member of a new class of models for high Reynolds number flows*, SIAM Journal on Numerical Analysis, vol. 58 (5), 2020, pp. 3068 – 3090.

[144] G. P. Galdi, An Introduction to the Mathematical Theory of the Navier–Stokes Equations, Springer Tracts in Natural Philosophy, Volume I, Springer-Verlag, New York, 1994.

[145] V. John, Reference values for drag and lift of a two-dimensional time-dependent flow around a cylinder, Numerical Methods in Fluids, 44(7), 777-788, 2004.

[146] J.-W. Bao and J. M. Wilczak and J.-K. Choi and L. H. Kantha, Numerical simulations of air-sea interaction under high wind conditions using a coupled model: A study of hurricane development, Monthly Weather Rev., 128, pp. 2190-2210, 2000.

[147] F. O. Bryan and B. G. Kauffman and W. G. Large and P. R. Gent, The NCAR CESM flux coupler, Tech. Report NCAR/TN-424+STR, National Center for Atmosphere Research, 1996.

[148] X. Xie and D. Wells and Z. Wang and T. Iliescu, Approximate deconvolution reduced order modeling, Computer Methods in Applied Mechanics and Engineering, vol. 313(1), pp. 512-534, 2017.

[149] W. Layton and R. Lewandowski, *A High Accuracy Leray-Deconvolution Model of Turbulence and its Limiting Behavior*, Analysis and Applications, vol.6(1), pp. 23–49, 2008.

[150] L. Berselli and T.-Y. Kim and L. Rebholz, *Analysis of a reduced-order approximate deconvolution model and its interpretation as a Navier–Stokes–Voigt regularization*, Discrete and Continuous Dynamical Systems, series B, 21(4), pp. 1027–1050, 2016.

[151] J. Mathew, *Large eddy simulation of a premixed flame with approximate deconvolution modeling*, Proceedings of the Combustion Institute, vol.29(2), pp. 1995–2000, 2002.

[152] M. Aggul and S. Kaya and A. Labovsky, *Two approaches to creating a turbulence model with increased temporal accuracy*, Applied Mathematics and Computation, vol.358, pp. 25–36, 2019.

[153] S. Stolz and N.A. Adams, An approximate deconvolution procedure for large-eddy simulation, in: *Physics of Fluids*, 11(7), 1699, 1999.

[154] O. Walsh, Eddy solutions of the Navier–Stokes equations, in: J.G. Heywood, K. Masuda, R. Rautmann, V.A. Solonnikov (Eds.), The NSE II - Theory and Numerical Methods, in: LNM, vol. 1530, Springer, 1992.

[155] L.C.Berselli, On the Large Eddy Simulation of the Taylor-Green vortex, J. Math. Fluid Mech., 7 (2005), pp. 164–191.

[156] W. Layton, On Taylor/Eddy solutions of Approximate Deconvolution Models of Turbulence, Applied Mathematics Letters, 24 (2011), pp. 23–26.

[157] C. Ross Ethier and D.A. Steinman, Exact Fully 3D Navier–Stokes Solutions for Benchmarking, International J. Numer. Methods Fluids 19 (1994), pp. 369–375.

[158] Volker John and Joachim Rang, Adaptive time step control for the incompressible Navier–Stokes equations, Comput. Methods Appl. Mech. Engrg. 199 (2010), pp. 514–524

[159] Volker John, Finite Element Methods for Incompressible Flow Problems, *Springer Series in Computational Mathematics*, Springer, 2016.

[160] V. John and A. Liakos, Time dependent flow across a step: the slip with friction boundary condition, Int. J. Numer. Meth. Fluids, 50 (2006), pp. 713–731.

[161] W. Layton, C. Manica, M. Neda and L. Rebholz, Numerical analysis of a high accuracy Leray-deconvolution model of turbulence, N.M.P.D.E, 24(2) (2008), pp. 555–582.

[162] A. Labovsky, *Energy Balance and Scale Truncation of the Approximate Deconvolution with Correction*, JMAA, Vol. 543, Issue 2, Part 2, 2025, 128929.

[163] R.H. Kraichnan, *Inertial-range transfer in two- and three-dimensional turbulence*, J. Fluid Mech. 47 (1971) 525.

[164] Y. Batugedara, A. Labovsky, *Approximate Deconvolution with Correction – a high fidelity model for Magnetohydrodynamic flows at high Reynolds and magnetic Reynolds numbers*, Computational Methods in Applied Mathematics, vol. 24, no. 1, 2024, pp. 1-20.

[165] Y. Batugedara, A. Labovsky, K. Schwiebert, *Validation of LES-C turbulence models*, Computer Methods in Applied Mechanics and Engineering, vol. 478(A), 2024, 116458.

[166] W. Layton, R. Lewandowski,*A simple and stable scale-similarity model for large Eddy simulation: Energy balance and existence of weak solutions*, Applied Mathematics Letters, vol. 16(8), 2003, pp. 1205–1209.

[167] S. Pope, *Turbulent Flows*, Cambridge Univ. Press, 2000.

[168] Y. Batugedara, A. Labovsky and K. Schwiebert, *Higher temporal accuracy for LES-C turbulence models*, Computer Methods in Applied Mechanics and Engineering, vol. 377, 2021, 113696.

[169] Schäfer M, Turek S., The benchmark problem "flow around a cylinder". In *Flow Simulation with High-Performance Computers II*, Hirschel EH (ed.). Notes on Numerical Fluid Mechanics, vol. 52. Vieweg: Braunschweig, 1996; pp. 547–566.

[170] D. Erkmen, S. Kaya and A. Cibik, *A second order decoupled penalty projection method based on deferred correction for MHD in Elsässer variable*, Journal of Computational and Applied Mathematics, Vol. 371, 2020.

[171] J.-G. Liu and R. Pego, *Stable discretization of magnetohydrodynamics in bounded domains*, Commun. Math. Sci., vol. 8(1), 2010, pp. 234–251.

[172] J. Liu and W. Wang, *Energy and helicity preserving schemes for hydro and magnetohydro-dynamic flows with symmetry*, J. Comput. Phys., vol. 200, 2004, pp. 8–33.

[173] R. Beekie, T. Buckmaster, and V. Vicol, *Weak solutions of ideal MHD which do not conserve magnetic helicity*, Ann. PDE, vol 6(1), 2020, article 1.

[174] R. E. Caflisch, I. Klapper, and G. Steele, *Remarks on singularities, dimension and energy dissipation for ideal hydrodynamics and MHD*, Comm. Math. Phys., vol. 184(2), 1997, pp. 443–455.

[175] P. G. Schmidt, *On a magnetohydrodynamic problem of Euler type*, J. Differential Equations, vol. 74(2), 1988, pp. 318–335.

[176] P. Secchi, *On the equations of ideal incompressible magnetohydrodynamics*, Rend. Sem. Mat. Univ. Padova, vol. 90, 1993, pp. 103–119.

[177] M. Aggul, A. Labovsky, E. Onal, K. Schwiebert, *Fluid-Fluid Interaction Problems at High Reynolds Numbers: Reducing the Modeling Error with LES-C*, SIAM Journal on Numerical Analysis, 61(2), 2023, pp. 707–732.

[178] M. Aggul, *A grad-div stabilized penalty projection algorithm for fluid-fluid interaction*, Applied Mathematics and Computation, 414, 126670, 2022.

[179] J. Connors, *An Ensemble-Based Conventional Turbulence Model for Fluid-Fluid Interaction*, IJNAM, Volume 15, Number 4-5, Pages 492–519, 2018.

[180] M. Aggul and A. Labovsky, *Approximate Deconvolution Models for Turbulent Fluid-Fluid Interaction Problems*, Computers & Mathematics with Applications, vol. 117, pp. 113-126, 2022.

[181] M. Aggul, A. Labovsky and K. Schwiebert, *NS-ω Model for Fluid-Fluid Interaction Problems at High Reynolds Numbers*, Computer Methods in Applied Mechanics and Engineering, vol. 395, 115052, 2022.

[182] W. Layton, C. Manica, M. Neda, L. Rebholz, *Numerical analysis and computational comparisons of the NS-alpha and NS-omega regularizations*, Computer Methods in Applied Mechanics and Engineering, vol. 199, issues 13–16, 2010, pp. 916 – 931.

[183] W. Layton, I. Stanculescu, C. Trenchea *Theory of the NS-ω Model: a Complement to the NS-α Model*, Communications on pure and Applied Analysis, vol. 10(6), 2011, pp. 1763 – 1777.

[184] M. Aggul and S. Kaya, *Defect-deferred correction method based on a subgrid artificial viscosity model for fluid-fluid interaction*, Applied Numerical Mathematics, 160, pp. 178-191, 2021.

[185] W. Bangerth, D. Davydov, T. Heister, L. Heltai, G. Kanschat, M. Kronbichler, M. Maier, B. Turcksin and D. Wells. *The deal.II Library, Version 8.4*, Journal of Numerical Mathematics, vol. 25, no. 3, pp. 135-141, 2016 `http://www.dealii.org`.

[186] M. Aggul, *Defect correcting extrapolation technique for Oseen and Navier–Stokes Flows*, Computers & Mathematics with Applications, vol. 80, is. 5, pp. 1029-1049, 2020.

[187] M. Aggul, F. G. Eroglu, A. Labovsky, S. Kaya, *A projection based variational multiscale method for a fluid–fluid interaction problem*, Computer Methods in Applied Mechanics and Engineering, 365, n. 112957, 2020.

Index